Sustainable Civil Infrastructures

Editor-in-chief

Hany Farouk Shehata, Cairo, Egypt

Advisory Board

Khalid M. ElZahaby, Giza, Egypt
Dar Hao Chen, Austin, USA

Sustainable Infrastructure impacts our well-being and day-to-day lives. The infrastructures we are building today will shape our lives tomorrow. The complex and diverse nature of the impacts due to weather extremes on transportation and civil infrastructures can be seen in our roadways, bridges, and buildings. Extreme summer temperatures, droughts, flash floods, and rising numbers of freeze-thaw cycles pose challenges for civil infrastructure and can endanger public safety. We constantly hear how civil infrastructures need constant attention, preservation, and upgrading. Such improvements and developments would obviously benefit from our desired book series that provide sustainable engineering materials and designs. The economic impact is huge and much research has been conducted worldwide. The future holds many opportunities, not only for researchers in a given country, but also for the worldwide field engineers who apply and implement these technologies. We believe that no approach can succeed if it does not unite the efforts of various engineering disciplines from all over the world under one umbrella to offer a beacon of modern solutions to the global infrastructure. Experts from the various engineering disciplines around the globe will participate in this series, including: Geotechnical, Geological, Geoscience, Petroleum, Structural, Transportation, Bridge, Infrastructure, Energy, Architectural, Chemical and Materials, and other related Engineering disciplines.

More information about this series at http://www.springer.com/series/15140

Sherif El-Badawy · Jan Valentin

Editors

Sustainable Solutions for Railways and Transportation Engineering

Proceedings of the 2nd GeoMEast International Congress and Exhibition on Sustainable Civil Infrastructures, Egypt 2018 – The Official International Congress of the Soil-Structure Interaction Group in Egypt (SSIGE)

 Springer

Editors
Sherif El-Badawy
Public Works Engineering Department,
   Faculty of Engineering
Mansoura University
Mansoura, Egypt

Jan Valentin
Department of Road Structures,
   Faculty of Civil Engineering
CTU in Prague
Prague, Czech Republic

ISSN 2366-3405   ISSN 2366-3413   (electronic)
Sustainable Civil Infrastructures
ISBN 978-3-030-01910-5   ISBN 978-3-030-01911-2   (eBook)
https://doi.org/10.1007/978-3-030-01911-2

Library of Congress Control Number: 2018957405

This Springer imprint is published by the registered company Springer Nature Switzerland AG
The registered company address is: Gewerbestrasse 11, 6330 Cham, Switzerland

Contents

About the Editors

Sherif El-Badawy Public Works Engineering Department, Faculty of Engineering, Mansoura University, Mansoura, Egypt

sbadawy@mans.edu.eg

Sherif El-Badawy is an associate professor at Mansoura University, Egypt. He serves as the director of Highway and Airport Engineering Laboratory, and director of the Center of Scientific, Experimental, and Technical Services, Mansoura University. He received a B.Sc. degree with honor in Civil Engineering and M.Sc. from Mansoura University, Egypt. He pursued his Ph.D. degree at Arizona State University (ASU), USA, in 2006. He worked as a postdoctoral research associate at ASU from June 2006 until July 2007 and as a research fellow at University of Idaho from December 2009 until January 2012. He has more than 20 years of experience in pavement structural analysis, design, and characterization. He has participated in several state and national projects during his employment in the USA. He was one of the team members under Professor Witczak in the NCHRP 1-37A and 1-40D projects which focused on the development of the Mechanistic-Empirical Pavement Design Guide (MEPDG). He serves as a TRB Committee member on Flexible Pavement Design (AFD60) since April 2012. He is an elected board member of the Middle East Society of Asphalt Technologists (MESAT) and a member of the International Geosynthetic Society (IGS), Transportation and Traffic Society (TTS), International

Association of Computer Science and Information Technology (IACSIT), and Chi Epsilon Honor Society. His research interests focus on pavement material characterization and modeling, mechanistic-empirical pavement design methods, and traffic characteristics. He was the chief editor of the ASCE GPS "Innovative Technologies for Severe Weathers and Climate Changes." Through research and graduate advising, he published more than 70 technical publications and reports.

Jan Valentin is active in civil engineering (focus on pavement and road engineering) for more than 15 years. Since 2004, he was employed by the Faculty of Civil Engineering, Czech Technical University, in Prague. Presently, he is the deputy head of the Department of Road Structures and a senior researcher. He is also working part time as a specialist for the international company HOCHTIEF. He obtained his Ph.D. degree from CTU in Prague in 2009, focusing in his Ph.D. thesis on cold recycling performance characterization in 2009.

His main areas of research and scientific interest are recycling, warm mix asphalts, modification of bitumen incl. rubber, performance testing of bitumen and asphalt mixtures focusing actually mainly on deformation behavior. He is further interested in PPP schemes and in introducing digitalization to the road construction industry in the Czech Republic.

He represents Czech Republic in the HSE Committee of European Asphalt Pavements Association, and he is a member of IASS and some TC within RILEM.

Estimation of Railway Track Longitudinal Profile Using Vehicle-Based Inertial Measurements

Paraic Quirke[1], Eugene J. OBrien[2(✉)], Cathal Bowe[3],
Abdollah Malekjafarian[2], and Daniel Cantero[4]

[1] Murphy Surveys, Kilcullen, Co. Kildare, Ireland
[2] School of Civil Engineering, University College Dublin, Dublin 4, Ireland
eugene.obrien@ucd.ie
[3] Technical Department, Engineering & New Works, Iarnród Éireann/Irish Rail,
Inchicore, Dublin 8, Ireland
[4] Department of Structural Engineering, Norwegian University of Science
and Technology, Trondheim, Norway

Abstract. This paper presents an optimization algorithm for finding the railway track longitudinal profile from the inertial response of a train bogie. The track profile is that which creates a numerical response from a numerical train-track interaction model that best fits the measured response. An Irish Rail InterCity train was instrumented to capture in-service vehicle responses to validate the proposed algorithm. Several train passes in the Dublin-Belfast service in Ireland over a period of 4 weeks are used. Vertical acceleration and angular velocity at the train bogie were measured. The real longitudinal profile of a section of this line was surveyed by traditional means, for reference. The vehicle properties were available from a calibration study. The track profile is estimated using the proposed method which matches quite well with the surveyed profile. The reproducibility of the method is assessed.

1 Introduction

Real-time automated monitoring systems are beneficial in terms of the effectiveness of condition monitoring in identifying maintenance issues in any structural health monitoring system (Malekjafarian et al. 2015). This is particularly important for railway track monitoring considering the value of the asset and the implications of track failure (Salvador et al. 2016). The current method of railway track monitoring in many countries is using track recording vehicles (TRVs) to measure geometric properties of railway track. It ensures the safety of tracks against a suitable standard (Kouroussis et al. 2011). However, using TRVs includes high running costs and the number of measurement operations undertaken annually is limited. Track faults might go undetected between TRV operations which means that the measured data may lack the frequency required to identify faults on time and may not enable statistical analysis to identify trends in fault progression and track deterioration. There is a need for more regular information on track condition which can be obtained by measuring the

© Springer Nature Switzerland AG 2019
S. El-Badawy and J. Valentin (Eds.): GeoMEast 2018, SUCI, pp. 1–6, 2019.
https://doi.org/10.1007/978-3-030-01911-2_1

response of in-service trains using low-cost inertial sensors (Weston et al. 2015). This has the potential for a higher rate of data which facilitates more efficient maintenance planning.

Recently, inertial sensors installed on in-service trains are used to determine individual elements of track geometry (Malekjafarian et al. 2018). Track features such as rail irregularities, corrugation, vertical alignment, track stiffness, weld depressions, and changes in rail bending properties due to the presence of welds, cracks or other defects influence the response measured on a passing train. Several studies employ frequency based analysis to differentiate short- and long-wave defects. Measurements taken from the bogie or car body can be used to determine long-wave features. Short-wave defects such as corrugation generally require high frequency measurements taken at the axle box (Ward et al. 2011).

This paper presents an optimization method for finding the railway track profile using a calibrated vehicle model. The method is an inverse technique in which acceleration and rotation measurements from an Irish Rail train are employed to find the longitudinal track profile. The method infers a loaded track profile considering both the track longitudinal profile and the variation in longitudinal track stiffness. Cross Entropy (CE) optimisation (Blossom 2006) is employed. CE is an iterative method which uses Monte Carlo simulation to generate a population of trial solutions from a mean and standard deviation for each variable being sought. The population is updated and improved in each iteration based on an analysis of the optimum results in the previous iteration. The method is particularly suited to multi-variate problems with large solution spaces as is typical in Civil Engineering (Quirke et al. 2017; OBrien et al. 2017).

2   Train Instrumentation

The acceleration response data was acquired using inertial sensors installed on the trailer (non-powered) bogie of an Irish Rail Hyundai Rotem InterCity fleet car. The instrumentation was installed in December 2015 on the leading car (22337) of Set 37, a 5-car train set. This set was configured for operation in both Northern Ireland and the Republic of Ireland to replace the Dublin-Belfast Enterprise service while it was being refurbished. This meant that the set would serve the route exclusively, providing the maximum line repeatability possible for any train in the Irish Rail fleet.

A tri-axial accelerometer and tri-axial gyrometer were installed as close to the bogie centre of mass as possible (see Fig. 1). The properties of the installed sensors are listed in Table 1. The measured data were sampled at a frequency of 500 Hz. Testing was carried out on the Dublin-Belfast line from 13th January to 3rd February, 2016. 57 return journeys were made on the line.

Fig. 1. Bogie mounted gyrometer and accelerometer

Table 1. Properties of sensors

Sensor location	Type	Name	Range
Bogie – Bounce	Triaxial accelerometer	Disynet-DA3802-015g	±15 g
Bogie – Pitch	Triaxial gyrometer	Crossbow VG400CC-200	±200 °/s

3 Methodology

The aim of this investigation is to find the longitudinal profile of a railway track from the measured inertial response of an in-service train. Measured data is used as input to a numerical optimisation method executed in Matlab. Multiple datasets are used to infer track longitudinal profiles though a section of track with a known settlement so that comparisons to surveyed track longitudinal profile can be made.

Cross Entropy (CE) optimisation is used in this study to find the track longitudinal profile from the measured inertial responses of an in-service train. The CE method uses Monte Carlo simulation to generate a population of trial solutions from a mean and standard deviation for each variable being sought. A population of trial track longitudinal profiles is generated to be used in the numerical train-track interaction model (Fig. 2). The properties of the vehicle are given in Table 2. The acceleration and rotation responses of the bogie are generated by running the model and are compared to measured data to find an 'elite' set of profiles, i.e. the set of profiles that generate acceleration signals that most closely match the measurements. Then the elite set is employed to improve the population of estimates until the method converges to the solution.

4 Results

Of the 57 runs of the train, there were 46 where data was acquired at a scan rate of 500 Hz. The optimization algorithm is applied these 46 datasets recorded in the experiment. Inferred longitudinal track profiles are filtered using a 6th order Butterworth band-pass filter between wavelengths of 3–25 m to match the D1 wavelength

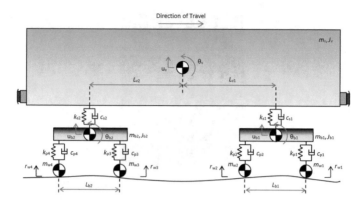

Fig. 2. 2D car model

Table 2. Properties of Irish Rail Hyundai-Rotem Intercity car

Property	Unit	Symbol	Value
Wheelset mass	kg	$m_{w1}, m_{w2}, m_{w3}, m_{w4}$	1 407
Bogie mass	kg	m_{b1}, m_{b2}	3 910
Car body mass	kg	m_v	36 852
Moment of inertia of bogie	kg.m^2	J_{b1}, J_{b2}	10 024
Moment of inertia of main body	kg.m^2	J_v	560 342
Primary suspension stiffness	N/m	$k_{p1}, k_{p2}, k_{p3}, k_{p4}$	2.8×10^6
Secondary suspension stiffness	N/m	k_{s1}, k_{s2}	1.0×10^6
Primary suspension damping	Ns/m	$c_{p1}, c_{p2}, c_{p3}, c_{p4}$	29.4×10^3
Secondary suspension damping	Ns/m	c_{s1}, c_{s2}	60×10^3
Distance between car body centre of mass and bogie pivot	m	L_{v1}, L_{v2}	8.0
Distance between axles	m	L_{b1}, L_{b2}	2.3

interval specified by the track geometry standard, EN 13848. A cross-correlation function is used to align the inferred profiles as the GPS lacks the accuracy and repeatability necessary for this task. Figure 3 shows the profiles inferred from the bogie inertial response data, measured through the case study section in 46 different runs. Aside from a surveyed dip at a track distance of 78.03 km, all other major features are detected. It can be seen that the magnitude of the elevation changes appear to be overestimated by the algorithm by a factor of about 2. Errors in the gyrometer calibration may be partially responsible for the poor match to the elevation profile in terms of magnitude. The use of a 2D vehicle model and possible inaccuracies in vehicle calibration, resulting in a poor representation of the vehicle dynamics, may also be a factor. Furthermore, the track was not loaded during the level survey, while the inertial measurements were taken on the train bogie and hence are from a loaded track. This may be a factor for the overestimation of the elevation profile where the inferred profile is less than the surveyed profile.

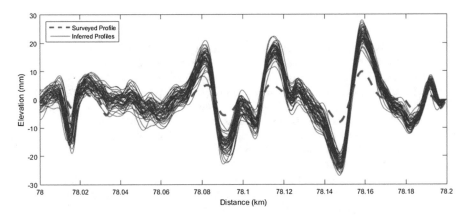

Fig. 3. Filtered inferred profiles, r(x), for 46 runs, and filtered surveyed profile (3–25 m)

5   Conclusions

An Irish Rail intercity train was instrumented for a period of 1 month to test a method for inferring railway track longitudinal profile through optimisation. Inertial sensors were placed on a non-powered trailer bogie of a coach at the front of the train. Bogie vertical acceleration and angular velocity are extracted through an area of known track settlement which was surveyed during the measuring period. Cross Entropy optimisation is used to find the track longitudinal profile that generates vehicle model responses that best match the measured signals. A reasonable match is achieved between the inferred longitudinal profiles and the surveyed track profile, albeit with an overestimation of the profile elevations. Better results would be expected if the comparison were made against loaded track as opposed to unloaded as here. Aside from reproducibility and scaling issues, the method presented in this chapter can be readily implemented using low-cost inertial sensors and simple numerical modelling. The method can give very frequent updates of the track profile allowing the identification of differences in profiles that emerged suddenly or over extended periods of time.

Acknowledgments. The research presented in this paper was carried out as part of the Marie Curie Initial Training Network (ITN) action FP7-PEOPLE-2013-ITN. The project has received funding from the European Union's Seventh Framework Programme for research, technological development and demonstration under grant agreement number 607524.

References

Blossom, P.: The cross-entropy method: a unified approach to combinatorial optimization, Monte-Carlo simulation, and machine learning. Interfaces **36**, 92–93 (2006)

Kouroussis, G., Verlinden, O., Conti, C.: Free field vibrations caused by high-speed lines: measurement and time domain simulation. Soil Dyn. Earthq. Eng. **31**, 692–707 (2011)

Malekjafarian, A., McGetrick, P.J., OBrien, E.J.: A review of indirect bridge monitoring using passing vehicles. Shock Vibr. (2015). http://dx.doi.org/10.1155/2015/286139

Malekjafarian, A., OBrien, E.J., Golpayegani, F.: Indirect monitoring of critical transport infrastructure: data analysis and signal processing. In: Alavi, A.H., Buttlar, W.G. (eds.) Data Analytics Applications for Smart Cities. Auerbach/CRC Press, USA (2018)

OBrien, E.J., Quirke, P., Bowe, C., Cantero, D.: Determination of railway track longitudinal profile using measured inertial response of an in-service railway vehicle. Struct. Health Monit. (2017). https://doi.org/10.1177/1475921717744479

Quirke, P., Cantero, D., OBrien, E.J., Bowe, C.: Drive-by detection of railway track stiffness variation using in-service vehicles. Proc. Inst. Mech. Eng. Part F J. Rail Rapid Transit **231**, 498–514 (2017)

Salvador, P., Naranjo, V., Insa, R., Teixeira, P.: Axlebox accelerations: their acquisition and time-frequency characterisation for railway track monitoring purposes. Measurement **82**, 301–312 (2016)

Ward, C.P., Weston, P.F., Stewart, E.J.C., Li, H., Goodall, R.M., Roberts, C., Mei, T.X., Charles, G., Dixon, R.: Condition monitoring opportunities using vehicle-based sensors. Proc. Inst. Mech. Eng. Part F J. Rail Rapid Transit **225**, 202–218 (2011)

Weston, P., Roberts, C. Yeo, G.: Perspectives on railway track geometry condition monitoring from in-service railway vehicles. Veh. Syst. Dyn. **53**, 1063–1091 (2015)

Soil Reinforcement Using Recycled Plastic Waste for Sustainable Pavements

Muhammad Hafez, Rabah Mousa, Ahmed Awed[(✉)],
and Sherif El-Badawy

Public Works Engineering Department, Faculty of Engineering, Mansoura
University, Mansoura, Egypt
{ammawed, sbadawy}@mans.edu.eg

Abstract. Resilient modulus (M_r) is a representative property for characterizing unbound granular materials and subgrade soils. It exhibits the elastic behavior as well as the load-bearing ability of pavement materials under cyclic traffic loads. This paper investigates the influence of using recycled plastic Polyethylene Terephthalate (PET) as a soil reinforcement material on the M_r of a clayey soil; ordinary soil in the delta region in Egypt. A comprehensive laboratory testing was conducted at Mansoura University Highway and Airport Engineering Laboratory (H&AE-LAB). The conducted testing includes standard engineering tests and repeated-loading triaxial tests (RLTT). Laboratory specimens were prepared at four different percentages of the recycled PET (0%, 0.2%, 0.6%, and 1.0%). RLTT results shows that the M_r of 0.6% PET-reinforced specimens increases by 58% compared to the M_r of the control specimen (0% PET). However, the M_r of the reinforced soil is found to decrease with the increase of PET percentage. Moreover, the universal M_r model exhibits excellent M_r predictions for the control and the PET-reinforced clay soil. Economically, the initial cost for constructing a 10-km road segment decreases by 8% using the 0.6% PET-reinforced Subgrade compared to the control Subgrade. Finally, damage analysis using the KENLAYER software is used to manifest the enhancement of pavement performance by reinforcing the Subgrade with PET.

Keywords: Resilient modulus · RLTT · PET · Soil reinforcement
Universal resilient modulus model · KENLAYER · Damage analysis
Pavement performance

1 Introduction

Soil properties have been improved through the historical backdrop of human civilization. With the present high demand of more sustainable structures, soil improvement procedures have turned into an essential task in the geotechnical engineering projects. Such strategies have been produced as per progresses in current innovation and human resources that make numerous structural building projects more practical (Patil et al. 2016a, b). Currently, different methods are used to enhance soil properties mechanically by compaction or by adding different types of materials that improve the characteristics of soils. Some of these materials that are being used as additives change

© Springer Nature Switzerland AG 2019
S. El-Badawy and J. Valentin (Eds.): GeoMEast 2018, SUCI, pp. 7–20, 2019.
https://doi.org/10.1007/978-3-030-01911-2_2

the chemical composition of soils such as cement and lime as. Other reinforcing materials are nonreactive such as fibers, geotextiles, and geogrids (Hejazi et al. 2012).

Soil reinforcement enhances the bearing capacity of the soil by compaction, proportioning and/or adding a suitable admixture. Reinforcing the subgrade soil which is the most vulnerable layer in the pavement structure improves its performance against various defects (Pundir and Prakash 2015). If soil enhancement gained such importance, recycling of plastics also is considered as one of the critical issues which has visibly aided in diverse parts of the life. The widespread reuse of plastics in many areas of public life in tandem with being non-biodegradable material raised the questions regarding its sustainability. Not to mention being one of the major problems for the environment (Bharadwaj et al. 2015). The global plastic waste increased dramatically in the past decades, while reports indicate that approximately 60 million tons of plastic water bottles are consumed every day in the United States, as well as 86% of plastic water bottles become garbage which ends up in landfills throughout the country (Food & Water Watch 2017). In this research, plastic water bottles were reused to improve the soil stiffness.

2 Literature Review

Soil reinforcement with different additives and various techniques is of special concern to many researchers. Many studies aimed at using recycled plastic waste as a reinforcing material to different types of soils and even hot mix asphalt (HMA). Most of these studies succeeded in improving the stiffness and strength of the soil. It is noteworthy to mention that the studies advocated the use of clay soil through their experimental work are limited.

(Benson and Khire 1994) conducted cyclic triaxial tests on sand reinforced with recycled High Density Poly Ethylene (HDPE) strips. Strips of three different shapes with specific dimensions were used and the plastic content was varied between 1 to 4% by weight. Their study showed that the maximum increase in the M_r was 35% at 3% plastic content compared to the nonreinforced system. They also found that the plastic strips enhanced the shear strength of sand. (Choudhary et al. 2010) studied the effect of adding randomly oriented HDPE strips with various aspect ratios ranged from 1 to 3 on sand. The results showed that reinforcing sand with HDPE strips improved its resistance to deformation and strength as the maximum California Bearing Ratio (CBR) value of the reinforced material was approximately 3 times the nonreinforced one. They also found that the reinforcement benefit increased with an increase in waste plastic strip content and length while the maximum CBR was obtained at plastic content of 4% and aspect ratio of 3. (Rawat and Kumar 2016) also conducted CBR tests on HDPE strips reinforced clayey silt soil. The tests produced similar conclusions to (Choudhary et al. 2010)'s study.

Studies were also carried out on PET as a reinforcing material (Consoli et al. 2002), (Babu and Chouksey 2011), (Manuel and Joseph 2014), (Patil et al. 2016a, b). (Consoli et al. 2002) studied the influence of reinforcing cemented and uncemented fine sand with reclaimed PET fibers. It was found that the fibers enhanced the stress-strain response of the cemented and uncemented sand and also found that the greatest improvements in triaxial shear strength and ductility were observed for the longer, 36 mm fiber. Whereas (Babu and Chouksey 2011) reported that the presence of PET chips in fine sand or red soil increased the unconfined strength with the maximum increase at 1% plastic content. (Manuel and Joseph 2014) evaluated the effect of using waste PET strips as a reinforcement material in kuttanad clay and it was found that the CBR value increased with the increase in the strip content whilst the unconfined compressive strength clearly increased. While (Patil et al. 2016a, b) who also studied mixing black cotton soil with plastic waste observed that the inclusion of PET strips improved the cohesive property of the soil and hence the bearing capacity increased apparently.

Further studies investigated the feasibility of using other plastic materials such as Low Density Polyethylene (LDPE) and Polypropylene to reinforce subgrade soils (Maher and Woods 1990), (Fletcher and Humphries 1991), (Cai et al. 2006), (Muntohar 2009), (Kumar et al. 2016). (Maher and Woods 1990) stated that the inclusion of plastic fibers increased the sandy soil rigidity. (Fletcher and Humphries 1991) performed CBR tests on fine sandy silt with different configurations of polypropylene fiber. The maximum increase in the CBR value was about 133% compared to the nonreinforced material while the longer fibers yielded higher CBR values. (Cai et al. 2006) found that the technique of fiber reinforced lime soil provided an increase in the material strength. (Muntohar 2009) reported that the utilization of polypropylene fibers together with lime and rice husk ash advanced the soil stiffness as well as the compressive strength, particularly at the longer length. (Kumar et al. 2016) conducted CBR and permeability tests on dune sand supported with LDPE strips. They found that the reinforced system had slight improvement in the CBR value while the permeability reduced with plastic content increase.

The literature studies showed the benefits of reinforcing the weak subgrade soils on the strength gained by reinforcement. These studies were mainly conducted on sandy soils while very limited number of studies addressed the effect of reinforcement using plastic waste on the clayey subgrade soils. The majority of these studies only investigated the basic properties of the tested materials such as CBR, shear strength, and permeability. Only fewer studies addressed the effect of such reinforcement on the fundamental strength property of the reinforced soil i.e. M_r. Finally, limited studies quantified the effect of waste plastic reinforcement on pavement performance as well as the reduced construction cost due to reinforcement. Thus, the main objectives of this study are (a) improving soil stiffness by using PET and determining M_r of soil at different percentages of PET, (b) studying the effect of using PET on the pavement performance using KENLAYER software, and finally (c) quantifying the economic value of using PET waste in road construction in Egypt.

3 Materials and Methods

3.1 Materials

3.1.1 PET

In this study, recycled water bottles were manually cut into small strips of 2.5 cm length and a width of 1.0 cm with an aspect ratio of 2.5. The size was selected to ensure that the strips could be smoothly handled and mixed into soil. Recycled PET strips from water bottles were reported to have a specific gravity of about 1.4 gm/cm^3 (Babu and Chouksey 2011). Herein, the plastic strips were mixed with the soil in different percentages (0%, 0.2%, 0.6%, 1%) by dry weight of soil.

3.1.2 Clayey Soil

The natural soil was sampled from a construction site in Al-Mahalla Al-kubra City in the Delta region, Egypt. The soil is classified as (A-7-5) according to the AASHTO soil classification system and (CL) according to the Unified Soil Classification System (USCS). Its specific gravity (G$_s$) is 2.65, the percent passing No. 200 US sieve is 51.7%, liquid limit (L.L = 44.5) and plastic limit (P.L = 16.9). A summary of the soil basic properties along with the test standard is shown in Table 1.

Table 1. Basic engineering properties of the nonreinforced clayey soil

Test		Standard	Clay soil
Percentage of fine, P#200, %			51.7
Specific gravity (G$_{se}$)		ASTM D 854	2.65
Atterberg limits	Liquid limit, LL, %	ASTM D 4318	44.5
	Plastic limit, PL, %	ASTM D 4318	16.9
	Plasticity index, PI, %	ASTM D 4318	27.6
Maximum dry density, MDD, gm/cm^3		ASTM D 1557	1.63
Optimum moisture content, OMC, %		ASTM D 1557	15.5
Soaked California Bearing Ratio, (CBR), %		ASTM D 1883	6.68
AASHTO classification		ASTM C 136	A-7-5
Unified Classification			CL

MDD = Maximum Dry Density; OMC = Optimum Moisture Content; CBR = California Bearing Ratio

3.2 M$_r$ Testing

Split steel molds 200 mm high and internal diameter of 100 mm, were used for the material compaction. All specimens were compacted in five layers at the Maximum Dry Density (MDD) and optimum moisture content according to the Modified Proctor compaction effort shown in Table 1. The final moisture content after each test was determined.

For specimens reinforced with PET, the PET strips were added to the subgrade soil and mixed with three different percentages of (0.2%, 0.6% and 1%) by weight of the dry soil and adequately distributed through specimen.

Generally for all specimens, after preparation of specimens they were rapped air tight and left for two days for curing at room temperature before testing. The final height, diameter and weight of each specimen were recorded before performing the test. The rubber membrane was stretched around the specimen by the membrane stretcher and then the membrane was sealed to the end caps by means of tight O-rings. Two external Linear Variable Differential Transducers (LVDTs) with stroke lengths of 5.08 mm, were mounted externally to the cell. The LVDTs are attached to two aluminum clamps. Sample preparation is shown in Fig. 1.

Fig. 1. Specimen preparation procedure for M_r testing

All specimens were subjected to repeated load in a triaxial cell using the UTM-25 (available in H&AE-LAB) in accordance with AASHTO T307-99-2012 test protocol. Combinations of applied repeated vertical stress and static confining pressure were applied over 15 sequences to characterize the vertical resilient strain response. The number of load repetitions is 100 cycles per sequence. The load shape was haversine with 0.2 s loading duration and 0.8 s as rest period. Every specimen was conditioned with a confining pressure 41.4 kPa and a deviator stress of 24.8 kPa for 500 cycles as per the test protocol.

4   Results and Analysis

4.1   Effect of Adding PET in Soil

So as to study the impact of PET content on the resilient behavior of the subgrade soils, the M_r was determined at the anticipated field stresses using the universal model. (Ji et al. 2014) reported that the anticipated field confining stress for subgrade is in the range of 6.9 to 13.8 kPa (1 to 2 psi) while the anticipated field deviator stress in subgrade ranges from 41.4 to 48.3 kPa (6 to7 psi). The M_r for soils reinforced with PET at various percentages of PET are appeared in Fig. 1. The M_r of the soil reinforced with PET increased with an increase in PET percentage until the point when it achieved an ideal Mr at around 0.6% PET. Beyond after a decrease is noticed with increasing the

PET content as shown in Fig. 2. The change of the M_r reached around 59% at a PET percentage of 0.6% compared to the nonreinforced soil.

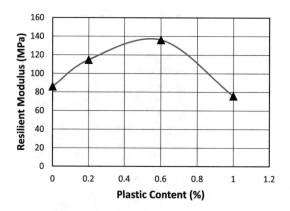

Fig. 2. The effect of PET content on the M_r of soil

4.2 M_r Modelling

The universal model is widely accepted models that is routinely used in different pavement design methods all over the world for modelling the M_r testing results. This model is used in this study to evaluate the impact of PET on the M_r and to assess the effect of soil reinforcement on the pavement behavior. This model is as follows:

$$M_r = K_1 * Pa * \left(\frac{\theta}{Pa}\right)^{K_2} * \left(\frac{\tau_{oct}}{Pa} + 1\right)^{K_3} \tag{1}$$

Where K_1, K_2, K_3 are regression coefficients, θ = bulk stress = $\sigma_1 + \sigma_2 + \sigma_3$, τ_{oct} is the octahedral shear stress = $\frac{\sqrt{2}}{3}[(\sigma_1 - \sigma_3)]$, Pa is the atmospheric pressure = 100 kPa, σ_1, σ_2, σ_3 = major, intermediate, and minor principal stress.

A nonlinear optimization technique (i.e., Solver in Microsoft Excel) was utilized to compute the regression constants of the model for the control and reinforced soils with the different percentages of PET. These coefficients along with the coefficient of determination are shown in Table 2. The M_r prediction accuracy in terms of coefficient of determination, R^2 for both unreinforced and reinforced subgrade soil was excellent, 99.51% for all types as shown in Fig. 3.

Table 2. The Universal Model constants for the investigated materials

	K_1	K_2	K_3	R^2	R^2_{ADJ}	SE/SY
Control soil	0.738	−0.021	0.467	0.86	0.84	0.37
0.2% PET	0.776	−0.015	1.239	0.98	0.97	0.15
0.6% PET	1.205	0.019	0.409	0.85	0.83	0.38
1% PET	0.682	0.036	0.368	0.52	0.44	0.69

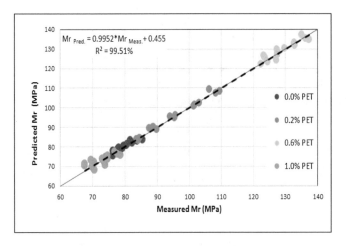

Fig. 3. Comparison of measured and predicted M_r using the universal model for all types

5   Performance Analysis Using KENLAYER

The KENLAYER software was utilized to perform damage analysis on a typical flexible pavement section with 10.16 cm (4 in.) asphalt layer atop 25.4 cm (10 in.) granular base layer resting on a subgrade layer as illustrated in Fig. 4. In addition to the virgin subgrade soil, the reinforced subgrade soil with 0.6% plastic content was studied while the asphalt and base layers properties were identical in the two cases. The pavement section was subjected to 80-kN (18-kips) equivalent single axle loads (ESALs) with 0.827 MPa (120 psi) tire pressure and 34.29 cm (13.5 in.) spacing between the dual tires.

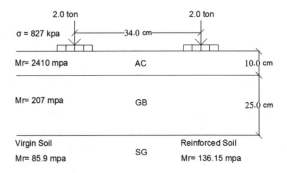

Fig. 4. Typical cross section used in the analysis

The M_r of fine grained soil decreases with the increase of deviator stress ($\sigma_d = \sigma_1 - \sigma_3$), stress softening. In layered system, σ_2 may not be equal to σ_3, so the average of σ_2 and σ_3 is considered as σ_3 including the layer system

$$\sigma_d = \sigma_1 - 0.5(\sigma_2 + \sigma_3) + \gamma_z(1 - k_\circ) \tag{2}$$

Equation (2) is not theoretically correct because the principal loading stress may not be in the same direction as the geostatic stresses. Because the loading stresses that reaches the subgrade layer are usually small, they do not have a significant effect on the computed modulus. The KENLAYER software uses three normal stresses σ_x, σ_y and σ_z to replace σ_1, σ_2 and σ_3 in the Eq. (2). Through the software if the point selected for computing the modulus is on the axis of symmetry for a single tire or on the plan of symmetry between dual tires, the three normal stresses and the three principal stresses are indicated (Huang 2004) (Fig. 5).

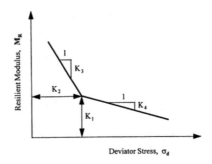

Fig. 5. The general relationship between M_r and σ_d

As shown in the Fig. 4 the bilinear behavior can be presented as following

$$M_r = K_1 + K_3(K_4 - \sigma_d) \text{ when } \sigma_d < K_2 \tag{3-a}$$

$$M_r = K_1 - K_4(\sigma_d - K_2) \text{ when } \sigma_d > K_2 \tag{3-b}$$

Where; K_1, K_2, K_3 and K_4 are the material constants. (Thompson and Elliott 1985) mentioned that the value of the M_r at the breakpoint in the bilinear curve is K_1 which is the good indicator of resilient behavior while K_2, K_3 and K_4 display the less variability and influence to the pavement response to a smaller degree than K_1. The soil in KENPAVE is classified into four types (very soft, soft, medium and stiff clay) with the M_r and deviator tress relationship. The maximum M_r is governed by a deviator stress 13.8 kPa however the minimum M_r is limited by the unconfined compressive strength which are assumed to be 42.8, 89.0, 157.0 and 226 kPa for four types of soils. Equation (3) has been inserted in KENLAYER software; however K_1 and K_2 herein were different from those used in (Thompson and Elliott 1985).

Table 3 discuss the inputs used in KENLAYER software for the section shown in Fig. 4 for both reinforced and unreinforced sections.

In this study, the vertical stress and vertical strain besides the horizontal strain at the critical sections were demonstrated to determine how far the plastic waste enhanced the pavement structure behavior against the heavy loads.

Table 3. The KENLAYER inputs for both reinforced and unreinforced sections

		Nonreinforced section	0.6%PET reinforced section
Type of material		Nonlinear	Nonlinear
No. of layers		Three	Three
Unit weight (ton/m^3)	**Asphalt**	2.32	2.32
	Base	2.16	2.16
	Subgrade	1.63	1.58
No. of Z coordinates		15	15
No. of X – Y coordinates		25	25
Nonseasonal input parameters			
K_2 (MPa)		0.037	0.033
K_3 (MPa)		0.678	0.98
K_4 (MPa)		1.279	2.243
K_o (MPa)		0.004	0.004
Mr_{min} (MPa)		7.429	120.276
Mr_{max} (MPa)		76.418	124.067
K_1 (MPa)		78.803	127.016

As highlighted in Fig. 6, the left side represents the vertical stress under the contact area of the wheel vs. depth using an nonreinforced subgrade while the other side has PET-Reinforced subgrade at 0.6% plastic content. The vertical stress under the contact area of the wheel vs. depth was almost identical.

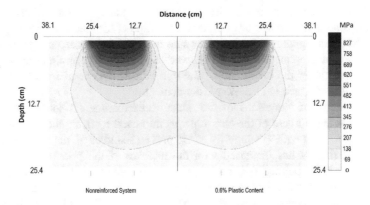

Fig. 6. Vertical stress distribution in the original and reinforced sections

In Fig. 7, the left side represents the vertical stress at the centerline of the wheel vs. depth using an nonreinforced subgrade while the other side has PET-Reinforced subgrade at 0.6% plastic content. The vertical stress at the centerline of the wheel vs. depth was the same in the two cases.

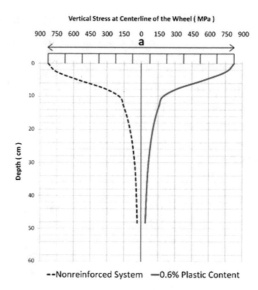

Fig. 7. Vertical stress vs depth under the centerline of the wheel in the original and reinforced sections

As shown in Fig. 8, the left side represents the horizontal tensile strain at the bottom of the asphalt layer vs. depth using an unreinforced subgrade while the other side has PET-Reinforced subgrade at 0.6% plastic content. The horizontal tensile strain at the bottom of the asphalt layer under the centerline of the wheel decreased by 3.25% through reinforcing the subgrade with 0.6% plastic content.

The left side represents the vertical compressive strain at the top of the subgrade layer, centerline of the asphalt layer and the centerline of the base layer vs. depth using an unreinforced subgrade while the other side has PET-Reinforced subgrade at 0.6% plastic content as presented in Fig. 9. The vertical compressive strain at the top of the subgrade layer under the centerline of the two wheels decreased by 12.7% through reinforcing the subgrade with 0.6% plastic content.

Thus, in light of the results of the conducted experimental study, the M_r tests suggested that the stiffness of the clay soil was improved by using the recycled plastic waste. In conformity with the majority of the previous studies on the different types of soil, the soil rigidity has increased with the increase of the plastic content. In this regard, it should be pointed out, however, the soil stiffness was increasing until the greatest increase at certain plastic content then the soil collapsed with any further increase in plastic content down to below the virgin soil results which could be interpreted as disintegration between the clay particles as a result of occupying a large volume of the specimen.

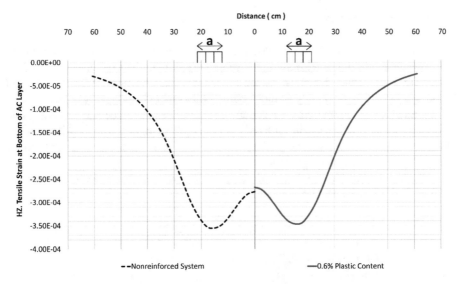

Fig. 8. Comparison of the horizontal tensile strain at bottom of AC layer of the original and reinforced sections

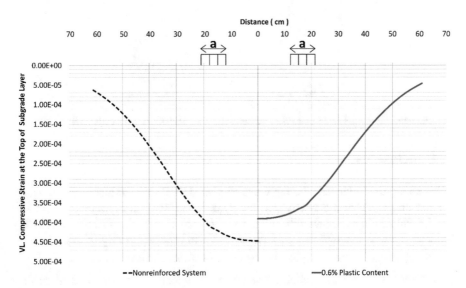

Fig. 9. Comparison of vertical strain at top of subgrade layer of the original and reinforced sections

6 Economic Study

In an attempt to keep the study applicable, an economic study was conducted to estimate the feasibility of reinforcing the subgrade soil with recycled plastic while (at a time when) the results depended on the latest local market prices of 2017. A typical cross section of a flexible pavement with 27.94 cm (11 in.) asphalt layer above 29.21 cm (11.5 in.) granular base layer and nonreinforced subgrade soil was compared to a section with reinforced soil for the top 30 cm of the subgrade using AASHTO1993 design method to measure the reduction in the layers thicknesses as a result of increasing M_r values of subgrade layer.

As evidenced in Fig. 10 that reinforcing the soil with recycled plastic reduced the total cost of the pavement construction by 8%.

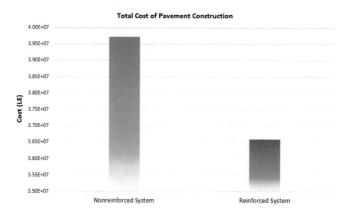

Fig. 10. Total cost of pavement construction for reinforced & nonreinforced systems

7 Conclusions

In this paper, the influence of using recycled plastic PET strips as a soil reinforcement material on the M_r of a clay soil was invitigated. The PET strips were cut in size of 1×2.5 cm and incorporated into soil at four portions of (0%, 0.2%, 0.6%, and 1.0%). The tests showed a significant enhancement due to reinforcing clay with PET strips. The test results obtained however led to the following conclusions:

1. Subgrade soil can be reinforced using recycled plastic bottles as an alternative method for improving the performance of the pavement structure.
2. Reinforcing subgrade soil with plastic waste increases its M_r until certain content of PET then the material weakens while the PET content increases.
3. The M_r values are well predicted with the universal model.
4. Reinforcing subgrade layer with 0.6% PET enhances the pavement structure performance through resisting both fatigue cracking and rutting.

The results of this study generated that recycled PET strips may improve the stiffness of clay soil. Nevertheless, further study is needed to manifest the optimal size of the strips with the maximum increase in the M_r value.

References

American Association of Highway and Transportation officials T307-99. Standard Method of Test for Determining the Resilient Modulus of Soil and Aggregate Materials (2003)

American Society for Testing and Materials (ASTM), Standard Test Method for Sieve Analysis of Fine and Coarse Aggregates 1, Designation C136-01 (2001)

American Society for Testing and Materials (ASTM), Standard test method for Laboratory Compaction Characteristics of Soil Using Modified Effort, Designation D 1557 (1995)

American Society for Testing and Materials (ASTM), Standard Test Method for California Bearing Ratio (CBR) of Laboratory-Compacted Soils, Designation D 1883 (1995)

American Society for Testing and Materials (ASTM), Standard Test Methods for Liquid Limit, Plastic Limit, and Plasticity Index of Soil, Designation D 4318 (1995)

Babu, G.L.S., Chouksey, S.K.: Stress-strain response of plastic waste mixed soil. Waste Manag. **31**(3), 481–488 (2011). https://doi.org/10.1016/j.wasman.2010.09.018

Benson, C.H., Khire, M.V.: Reinforcing sand with strips of reclaimed high-density polyethylene. J. Geotech. Eng. Am. Soc. Civ. Eng. **120**(5), 838–855 (1994)

Bharadwaj, A., Yadav, D., Varshney, S.: Non-biodegradable waste – its impact, safe disposal, pp. 391–398 (2015)

Cai, Y., et al.: Effect of polypropylene fibre and lime admixture on engineering properties of clayey soil. Eng. Geol. **87**(3–4), 230–240 (2006). https://doi.org/10.1016/j.enggeo.2006.07.007

Choudhary, A.K., Jha, J.N., Gill, K.S.: A study on CBR behavior of waste plastic strip reinforced soil. Emir. J. Eng. Res. **15**(1), 51–57 (2010)

Consoli, N.C., et al.: Engineering behavior of a sand reinforced with plastic waste. J. Geotech. Geoenviron. Eng. **128**(6), 462–472 (2002). https://doi.org/10.1061/(asce)1090-0241(2002) 128:6(462)

Fletcher, C.S., Humphries, W.K.: California bearing ratio improvement of remolded soils by the addition of polypropylene fiber reinforcement. Transp. Res. Rec. **1295**, 80–86 (1991). http:// trid.trb.org/view.aspx?id=359118

Food & Water Watch: How the Fracking Industry Profits off of Bottled Water (2017). https:// www.foodandwaterwatch.org

Huang, Y.H.: Pavement Analysis and Design, 2nd edn. Pearson/Prentice Hall, Upper Saddle River (2004)

Kumar, P., et al.: Open access stabilization of dune sand mixed with plastic (LDPE) waste strips for design of flexible pavement in construction of roads. Am. J. Eng. Res. **5**(11), 315–320 (2016)

Maher, B.M.H., Woods, R.D.: Dynamic Dynamic response of sand reinforced with randomly distributed fibers exhibit a relatively higher increase in resistance to liquefaction test equipment the shear modulus, G, and damping ratio, D, of RDFS composites were measured by both resonant-column and torsional shear. J. Geotech. Eng. **116**(7), 1116–1131 (1990)

Manuel, M., Joseph, S.A.: Stability analysis of kuttanad clay reinforced with PET bottle strips. Int. J. Eng. Res. Technol. **3**(11), 361–363 (2014)

Muntohar, A.S.: Influence of plastic waste fibers on the strength of lime-rice. Civ. Eng. Dimens. **11**(1), 32–40 (2009). http://puslit2.petra.ac.id/ejournal/index.php/civ/article/view/17028

Patil, A., et al.: Experimental review for utilisation of waste plastic bottles in soil improvement techniques. Int. J. Eng. Res. Appl. **6**(8), 25–31(7) (2016a)

Patil, P., et al.: Soil reinforcement techniques. Int. J. Eng. Res. Appl. **6**(8), 25–31 (2016b)

Pundir, V.S., Prakash, V.: Effect of soil stabilizers on the structural design of flexible pavements. J. Adv. Appl. Sci. Res. **6**(8), 134–147 (2015)

Hejazi, S.M., et al.: A simple review of soil reinforcement by using natural and synthetic fibers. Constr. Build. Mater. **30**, 100–116 (2012). https://doi.org/10.1016/j.conbuildmat.2011.11.045

Rawat, P., Kumar, A.: Study of CBR Behaviour of Soil Reinforced Hdpe, pp. 15–18, December 2016

Thompson, M.R., Elliott, R.P.: ILLI-PAVE-based response algorithms for design of conventional flexible pavements. Transp. Res. Rec. **1043**, 50–57 (1985)

Overview of Soil Stabilization Methods in Road Construction

Talal S. Amhadi[✉] and Gabriel J. Assaf

Department of Civil Engineering and Construction, École de Technologie
Supérieure (ÉTS), University of Québec, 1100 Rue Notre Dame Ouest, Montréal,
QC H3C, Canada
Talal.amhadi.1@ens.etsmtl.ca, Gabriel.Assaf@etsmtl.ca

Abstract. Soil stabilization is the process of improving the shear strength
parameters of soil and thus increasing its bearing capacity in road construction.
It is required when the soil available for construction is not suitable to carry
structural load. Generally, soils exhibit undesirable engineering properties
unless they are treated to enhance their physical properties. Stabilization can
increase the shear strength of a soil and control its shrink-swell properties,
thereby improving the load bearing capacity of a sub-grade to support pavement
and its foundations. Soil stabilization is used to reduce permeability and compressibility of the soil mass in earth structures and to increase its shear strength.
Mixing additives into the reaction mechanism, positively affecting its strength,
improving and maintaining the soil moisture content, achieve stabilization.
Therefore, these soil stabilization processes are suggested for most construction
systems and can be accomplished by several methods. All these methods fall
into two broad categories, namely mechanical stabilization and chemical stabilization. Mechanical Stabilization is the process of improving the properties of
the soil by changing its gradation; chemical stabilization is the process of adding
a physico-synthetic substance to the soil which reacts with the clay particles to
fill the voids so that less water is needed to maintain a stable mix and, finally, a
stable framework.

Keywords: Soil stabilization · Mechanical stabilization
Chemical stabilization · Strength

1 Introduction

To prepare soil as a base for road construction, a process of soil stabilization must be
undertaken to alter the soil properties, thereby improving the engineering qualities in
terms of workability, stiffness, permeability, strength, and compressibility. In recent
years, the practice of mixing cement into the soil has been adopted as a means of soil
stabilization in asphalt road construction; this has become a reliable method due to the
sizeable improvements that are observed in the properties of the treated soil. This
technique is most important for the base or subbase layers and it shows significant
advantages over a simple granular base with similar axel loads. There are a few options
to attain this stabilization; the most common is when the natural soil is mechanically
mixed with the stabilizing material to finally create a homogeneous mix (e.g. of cement

© Springer Nature Switzerland AG 2019
S. El-Badawy and J. Valentin (Eds.): GeoMEast 2018, SUCI, pp. 21–33, 2019.
https://doi.org/10.1007/978-3-030-01911-2_3

and natural soil); another option is to add the stabilizing material (e.g. cement or lime) to the surface of the soil and using compaction methods to make the stabilizing material penetrate into the voids in the soil (Bandara 2015). The use of stabilizing materials helps achieve particle cohesion in the soil, help achieve the desired moisture content of the soil, and help in the waterproofing and cementing of the soils (Addo et al. 2004). When the subgrade consists of clay soils, this creates obstacles for civil engineering because such soils tend to expand with increasing moisture content and this affects the service life of the road (Okasha and Abduljauwad 1992). Researchers have experimented with different kinds and quantities of stabilizing agents and have found that, for roadworks, lime and cement are some of the best options. Most often, these are added to the subbase and subgrade of the road, but it is also effective to add these to the base layers of the road construction (O'Flaherty 2002). For roads, there are many soils or granular materials available for road construction, but they may have inadequate properties (e.g., low load-bearing capacity, susceptibility to frost damage, etc.), which results in a significant roadside disaster and reduction of the pavement life. However, the addition of a stabilizing agent can improve soil properties. The most common additives are fly ash, lime, cement, and bitumen; other, less traditional options include liquid polymers, acids, silicates, lignin derivatives, resins, ions, and enzymes. Of these, the cement-treated base (CTB) results in a high degree of stiffness with the resulting high pavement durability. This research extends back to 1917 and there has been a significant publication history since then (Baghini et al. 2017).

The main objectives of this research study are addressed to different types of stabilization and comparing the advantage and disadvantage of each method of stabilization.

2   Types of Soil

The soil is a blend of organic matter, minerals, liquids, gases and countless organisms that together support life on Earth. Soil is constantly changing due to any number of chemical, biological, and physical processes, including climatic weathering, which involves erosion by wind and rain. Soft soils are most in need of stabilization if they are to be used in an engineering capacity; these soils include organic soils, and soils with notable amounts of peat, silt, or clay. Generally, the easiest soils to stabilize are fine-grained and granular because they have a relatively large surface area as compared to the diameter of the particle. Due to its elongated and flat shaped particles, clay soils have larger surface area relative to its particle size (Rogers et al. 1993). Some materials, silty soils are more difficult to stabilize because they are very sensitive to even the smallest variation in levels of moisture (Sherwood 1993). Organic soils, as well as soils with high peat content, can have very high water levels (as much as 2000%) and are highly porous. For example, peat soils can range from very muddy to quite fibrous; generally, peat soils have a shallow deposit but can, in some cases, go several meters deep (Pousette et al. 1999) and (Al Tabbaa and Stegemann 2005). Soil with high levels of organic content can also exchange large amounts of moisture, which can interfere with hydration because the soil keeps the calcium ions that are freed when the and calcium aluminate and the calcium silicate are hydrated. Therefore, in these organic soils, stabilization is only effective with the correct binder selection and application (Hebib and Farrell 1999).

3   Stabilization of Soils

Kearney and Huffman (1999) noted that the primary cause of premature asphalt pavement failure is because of unsuitable design, construction, and materials, with regard to environmental conditions, etc. Adding an agent to stabilize the soil is a means of improving the soil from an engineering point of view. Such a chemical means of stabilization using a manufactured product is only effective if it is added to the appropriate soil type and soil layer in the right amounts. Chemical agents for this purpose include fly ash, lime-cement, lime, Portland cement, or bitumen; these can be added individually or mixed with other agents (Olarewaju et al. 2011). Each stabilization agent has its own distinctive properties and will react with different soils differently depending on these properties and the properties of the soils. In Table 1, below, there is an overview of the different mechanisms and character of different stabilizing agents. It can be seen in the table that bitumen and cement are the best options for stabilizing non-plastic and granular soils; in contrast, for cohesive soils, lime is a better choice. Another case is the stabilization of granular soils; in this case it is more effective to add a coarse material (e.g. crushed gravel) to a fine material (e.g. sand).

4   Advantages and Disadvantages of Soil Stabilization

Many materials using as additives for Soil stabilization involve advantages and disadvantages. Some of the advantages and disadvantages of additives materials are discussed here.

5   Advantages of Soil Stabilization

By stabilizing the soil at the site of the construction, projects avoid the costs of removing the soil already there and transporting new materials to the site. In areas where extreme weather conditions would slow or stop construction during certain times of the year, soil stabilization can allow work to continue by stabilizing the original soil and allowing the work to continue. Therefore, stabilization techniques are a means of cost savings because work can continue through more weather conditions (Patel and Patel 2012).

The specific advantages of the treated soils are speeds up prior to the construction process since the required is usually much smaller and therefore less material and labor is required. Significantly improves strength and durability, especially where the local materials available soil is poor. May reduce or eliminate the need for the expensive surface treatment or rendering (Halland et al. 2012).

Table 1. Mechanisms and applicability of various stabilizing agents (Firoozi et al. 2017) and (Grogan et al. 1999).

Mechanism	Effects	Suitable soils
Granular Blending to poorly graded soils, usually coarse into fine (not clayey) soils	Higher compacted density, more uniform mixing, increased shear strength	Gap-graded or gravel deficient (gravel, sand addition), or harsh [a]FCR (loam addition)
Cement Mixing small amounts (cement modification) or larger proportions (cement binding) into soil or [a]FCR	Improve shear strength, reduces moisture sensitivity (modification), greatly increases tensile strength and stiffness (binding)	Most soils, especially granular ones, large amounts of cement needed in clay-rich and poorly graded sands, hence expensive
Lime Mixing hydrated lime or quick lime in small to moderate amounts into soils	Increases bearing capacity, dries wet soil, improves friability, reduces shrinkage	Cohesive soils, especially wet, high – PI clays
Lime Pozzolan Mixing lime plus fly ash or granulated slay into soil or [a]FCR	Similar to cement but slower acting and less ultimate strength	As for cement, plus clayey soils that do not react with lime
Bitumen Agglomeration, coating and binding of granular particles	Water proofs, imparts cohesion and stiffness	Granular, non-cohesive soils in hot climates
Fly ash Mixing with an activator to form cementitious compounds	Waterproof concrete	Some materials will activate the fly ash; lime or cement may be used to act as an activator by providing the required calcium hydroxide
Fiber The use of hair-sized polypropylene fibers in soil stabilization	Increases the stiffness of soil and also the immediate settlement of soil reduced considerably, the strength and angle of internal friction increase	Tropical soil, clay soil

6 Disadvantages of Soil Stabilization

The previous studies indicated the advantages of soil mixture. However, a number of disadvantages that are inherent in the treated soil, which can be identified as necessary stabilizing materials, may not be available in some developing countries or may be expensive for transportation. The mixing and building processes can be complicated depending on the type of stabilizer chosen. This can increase the likelihood of problems, which will affect the budget and time (Jawad et al. 2014). The deleterious chemical reactions there are two undesirable (deleterious) chemical reactions probably occur in the treated soil. The first is the carbonation and the second is the reaction with the sulfate salt existing in the soil. Carbonation is the reaction that occurs between the additives and atmospheric carbon dioxide (Umesha et al. 2009). According to (Cizer et al. 2006), the factors that controlling carbonation reaction are carbon dioxide diffusion through pores, calcium hydroxide and carbon dioxide dissolution in water, as well as the reaction of Ca+2 with CO3-2 ions to form the CaCO3 crystals.

7   The Process of Soil Stabilization

Proper design and testing are an important component of any stabilization project. Laboratory tests and on-site tests can establish proper design criteria when determining the appropriate rate for the addition of additives and impurities to achieve the desired engineering properties (Cortellazzo and Cola 1999). Stabilization of the soil is carried out in such a way that the stabilizing materials are distributed onto or mixed into the materials in need of stabilization. Generally, the additives are mixed into the soil until the desired properties are achieved, as shown in Fig. 1. Then the foundation or road materials are put in place. This process can vary depending on the necessary soils and additives (Cortellazzo and Cola 1999). In addition, it should be noted that the presence of sulfates, sulfides, carbon dioxide and organic substances in the stabilizing materials can contribute to unexpected or undesirable properties of the treated soil.

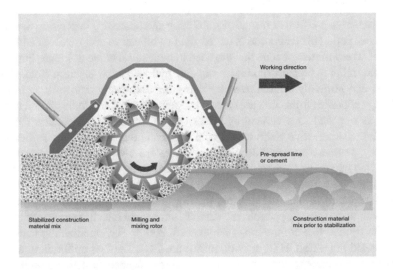

Fig. 1.   Schematic soil stabilization process (Cortellazzo and Cola 1999).

8   Techniques for Soil Stabilization

Soil or coarse aggregate is often used in roadworks, to build up the mass of materials needed to support the layers of the road. These soils and aggregates need to have very specific qualities to withstand the numerous and variable forces that they will undergo with axel-loads in an in-service road. Loose materials must be stabilized with some kind of binding agent, such as bitumen, lime, fly ash, cement or some mix of these. The end results will have greater strength, less compressibility and permeability than the untreated soil (Keller 2011). The engineering characteristics that are most important are durability, compressibility, permeability, strength, and volume stability (Sherwood 1993) and (Al Tabbaa and Stegemann 2005). The following are some reliable stabilization techniques such as mechanical stabilization, lime stabilization, cement stabilization, fly ash stabilization, bituminous stabilization, thermal stabilization, electrical stabilization, stabilization by geo-textile and fabrics, recycled and waste products etc.

8.1 Mechanical Stabilization

Mechanical stabilization is the oldest technique for road building and these involve changing the physically property of the original road base soil to affect its solidity and gradation, among other factors. Of these techniques, dynamic compaction is one of the most commonly used. To do this, a heavy weight is repeatedly dropped over all points of the road base to even out any irregularities and create an evenly compacted soil. A newer variation of this is known as Vibro compaction and it works in a similar way, with vibration taking the place of kinetic force deformation (Das 2003).

8.2 Lime Stabilization

Using lime is a cost-effective means of stabilizing and otherwise improving the material properties of soil. This technique is called lime stabilization and involves adding lime to the soil. The varieties of lime that are used to treat soil include dolomite lime, calcite quick lime, monohydrated dolomite lime, and hydrated high calcium lime. Soil stabilization can normally be achieved by using approximately 5% to 10% lime. The stabilizing properties from lime are caused by strengthening due to the cation exchange capacity as opposed to a cementing effect due to a pozzolanic reaction (Sherwood 1993). The most effective stabilization with lime can be done when the soil contains a lot of clay particles, which flocculate and therefore change natural plate-like clay particles into needle-like shapes with metalline structures that interlock. With the lime treatment, the clay soils become drier and are less affected by changes due to water content fluctuation (Keller 2011). Lime stabilization often creates pozzolanic reactions where, with added water, pozzolana materials have a chemical reaction with the lime and create cement-like compounds (White et al. 2005). The stabilization can be done with hydrated lime, $Ca(OH)_2$, or with quicklime, CaO. Another option is to use slurry lime when there are dry soils; in this case, water is needed to achieve optimal compaction (Rogers and Glendinning 1996). The most effective and common type of lime is Quicklime and its advantages, when compared with hydrated lime, are: it has a greater (per unit mass) free lime content; it has more density than hydrated lime (as such it requires less space for storage); it is generally less dusty; it generates some heat when it is mixed, a byproduct that adds to the strength of the final result; there is also a sizeable reduction in moisture levels, something that can be calculated with the following reaction equation:

$$CaO + H2O \rightarrow Ca(OH)2 + Heat\,(65\,kJ/mol) \tag{1}$$

Once mixed into soils with high moisture content, Quicklime absorbs moisture from the soils (as much as 32% of its weight), and it then forms hydrated lime. This reaction generates heat, which in turn causes moisture in the soil to evaporate; the process of the soil drying out and absorbing water is increased and is known as the soil's plastic limit (Sherwood 1993) and (Al Tabbaa and Stegemann 2005).

8.3 Cement Stabilization

The underlying mechanism in cement-like soil stabilization is that hydration of the particles creates interlocking crystalline bonds resulting in an elevated compressive strength. A successful bond results from the cement particles coating the majority of the soil particles. It is necessary to mix the cement with a soil that has a particular distribution of particle sizes if good stabilization is to be had (i.e. achieve good contact between the cement and soil particles). The resulting mix is an extremely compacted mix of cement and soil, sometimes referred to as a "cement-stabilized base" and sometimes a "cement-treated aggregate base". When the cement undergoes hydration, the resulting material is hard and long-lasting. This process of cement stabilization must be done during the compaction process. The void ratio of the soil is reduced as the cement takes up the void interspersed between the soil particles. Following this, if the soil takes on more water, the cement will react with the water and the mix becomes hard, causing the unit weight of the soil to increase. Through this process of hardening, the bearing capacity and the shear strength of the cement also increases. The effect that cement has on the clay-rich soils is that it decreases the soil's liquid limit and at the same time it increases the soil's workability and its plasticity index. The soil minerals do not affect the cement reaction but the presence of water will affect the cement (EuroSoilStab. 2002). For this reason, cement is useful for soil stabilization across a broad range of soil types. There are a great number of cement types on the markets including high alumina cement, sulfate resistant cement, blast furnace cement and standard Portland cement. The choice of which cement to use is a question of soil type and the required strength of the stabilization.

The process of hydration is the process that causes the cement to harden. The process starts when the cement is mixed with other components and water for the desired application, which leads to its hardening. The solidification of the cement will cover the surface of the soil as a glue coating, but it will not change the soil structure (EuroSoilStab. 2002). The hydration reaction proceeds slowly from the surface of the cement grains, and the center of the grains may remain non-hydrated (Sherwood 1993). Hydration of cement is a complex process with a complex series of unknown chemical reactions (Hicks 2002). However, this process can be influenced by what impurities or foreign agents are in the soil, the curing temperature of the cement, various possible additives in the cement, the ratio of water to the cement, and the characteristics of the mixture surface.

The final result and strength of the soil stabilized by cement depends on various factors and should be estimated during planning so as to attain the targeted strength. The two primary cementitious properties in standard Portland cement are C3S and C2S, two calcium silicates that undergird its strength (Hebib and Farrell 1999) and (MacLaren and White 2003). Another product of the hydration of Portland cement is calcium hydroxide. It reacts with five pozzolanic materials found in cement stabilized soil; this results in more cementitious materials (Cortellazzo and Cola 1999). Typically, the necessary quantity of cement is small but still enough to augment the soil's engineering properties and improve the cation exchange with the clay particles. The typical properties of soils stabilized with cement include a decrease in the range of expansion or compression volume, a decrease in the plasticity (i.e. in the cohesiveness), and an increase in the strength of the soil.

8.4   Stabilization with Fly Ash

Stabilization with fly ash is becoming more common and important in recent years, due possibly to its low expense and quick application, compared with other techniques. Fly ash has been used as a material for engineering for a long time, including many successful geotechnical implementations. As a byproduct of coal-fired electricity generation, fly ash has limited cementations properties when contrasted with cement or lime. The various kinds of fly ash are of use as secondary binders that do not exhibit cementitious effects by themselves. Nonetheless, when mixed with small quantities of activators, they can create chemical reactions that form cementitious mixes, helping to improve the strength of fine soils. The limitations of fly ash soil stabilization include the fact that fly ash and soil mixes that are cured at subzero temperatures and then is inundated with water are prone to strength loss and slaking (MacLaren and White 2003). Other limitations are that the target soil must have a much lower water content than with other methods (therefore dehydration may be needed prior to treatment). Finally, the sulfur content of the fly ash may bond with other elements in the mixture and form other minerals, thereby reducing the strength and lifespan of the stabilized soil.

8.5   Bituminous Stabilization

Bituminous soil stabilization is a method by which a controlled amount of bituminous material is thoroughly mixed with aggregate or an existing soil to form a wear surface or stable base. Bitumen increases the adhesion and bearing capacity of the soil and makes it resistant to the actions of water. Bitumen stabilization is carried out using asphalt, or asphalt emulsions. The bitumen type used depends on the kind of soil stabilized, the weather conditions, and the method of construction. In cold climates, the use of tar as a binder should be avoided because of its high-temperature maximum susceptibility. Tars and asphalts are bituminous materials that are used to stabilize the soil, usually for construction. Bituminous materials added to the soil impart cohesion and reduce water absorption (Kowalski and Starry Jr. 2007).

8.6   Thermal Stabilization

Thermal changes cause marked alterations in soil properties. Thermal stabilization is carried out either by heating the soil or by cooling it. An example of heating is as follows. As the soil heats up, its water content decreases. Electric repulsion between clay particles is decreased and the strength of the soil is increased. Freezing has a markedly different effect where cooling causes a slight loss of strength in the clay soils due to the increased repulsion of the particles. However, if the temperature drops below freezing, the soil stabilizes, and the porous water freezes (Lim 1983).

8.7 Electrical Stabilization

Electric stabilization of clay soils is carried out by a method known as electro-osmosis. When direct current (DC) is passed through clay soil, the porous water migrates to the negative electrode (cathode). This is due to the attraction of positive ions (cations) that are present in the water as they move towards the cathode. The strength of the soil is greatly increased by the removal of water. Electro-osmosis is an expensive method and is mainly used to drain cohesive soils. At the same time, the soil properties also improve.

8.8 Stabilization by Geo-Textile and Fabrics

Past studies have shown that the load capacity of subgrades and base course materials and their strength can be improved by incorporating non-biodegradable reinforcing materials such as fibers, geocomposites, geogrids and geotextiles. Geotextiles are porous fabrics made from synthetic materials such as polyester, polyvinyl chloride, nylons, and polyethylene. Varieties of mesh, woven, and nonwoven geotextiles are available. This method results in high strength stabilization. When it is properly introduced into the soil, it contributes to its stability. It is used in the construction of dirt roads on soft soils. A further means of strengthening the soil for stabilization is with metal strips woven into the geotextile, providing an anchor or a tie to restrain the skin cladding element of the geotextile (EuroSoilStab. 2002). These materials can be used to improve the durability and productivity of future highways and can reduce construction costs. At present, most of the research on these materials is based on tests conducted in the laboratory, which are only partially completed. Further laboratory tests and assessments will be required to develop design specifications based on the properties of the materials, and these specifications will need to be checked using large-scale field trials.

8.9 Recycled and Waste Products

For waste materials such as old crushed asphalt, copper and zinc slag, paper mill slag and rubber tires, improved methods of chemical and mechanical stabilization are needed. Because it is necessary to process many potentially hazardous materials, it will be necessary to develop realistic, economical and effective ways to assess the risk of contamination of these materials by leaching and emissions. In some cases, risk assessment is complicated by environmental regulations, and this issue must also be addressed.

9 Factors Influencing Soil Stabilization Strength

Organic materials, including, sulphides, sulphates, and carbon dioxide can affect the strength of soil stabilization (Sherwood 1993).

9.1 Organic Materials

Top layers of soil often contain a lot of organic material. In areas with good drainage, this organic material layer can be as deep as 1.5 m (Sherwood 1993). Organic materials in soil can react with products that stimulate hydration, for example calcium hydroxide (Ca(OH) 2). This lowers the pH of the soil. This lower pH can interfere with the 10 hydration process reduces the hardening of the stabilized soils, creating difficulties for compaction or making it completely impossible.

9.2 Sulphates

The use of a calcium-based stabilizer in sulphate-rich soils results in the stabilized sulphated soil reacting in the presence of excess moisture to form calcium sulphoaluminate (ettringite) or tamausite, a product that takes up more volume than the combined volume of reagents. However, in order to dissolve the sulphate, an excess of water is required for the mixing process. So that the reaction proceeds (Sherwood 1993) and (White et al. 2005).

9.3 Sulphides

Iron pyrites (FeS_2) sometimes occur in industrial byproducts and waste materials. When FeS_2 is oxidized, it produces sulphuric acid (H_2SO_4); when H_2SO_4 is mixed with calcium carbonate ($CaCO_3$), it can form gypsum (hydrated calcium sulphate) as seen in the following reactions:

$$2FeS_2 + 2H_2O + 7O_2 = 2FeSO_4 + 2H_2SO_4 \tag{2}$$

$$CaCO_3 + H_2SO_4 + H_2O = CaSO_4.2H_2O + CO_2 \tag{3}$$

Hydrated sulphate is the product of these reactions and, with excess water, it may undermine the soil stabilization, much lie sulphate would. That being said, gypsum also occurs in natural soil (Sherwood 1993) and (Al Tabbaa and Stegemann 2005).

9.4 Compaction

It is of considerable importance to consider the effect the binder creates with regard to the soil density. For any particular level of compaction, a stabilized mix will exhibit a maximum dry density that is lower than in unstabilized soil. With an increase in the amount of binders in the soil, the optimum moisture content also increases (Sherwood 1993). For example, in soils that have been stabilized with cement, the process of hydration occurs directly after cement is put in contact with the water. Because this instigates a hardening of the soil mix, it is critical to complete the soil compaction as soon as it can be done. In the case of delays, the compaction may have to account for the extra effort needed to compact a soil that has hardened due to the chemical reactions. Depending on the stage of the reactions, such mechanical compaction could result in significant breakdown of chemical bond the resulting strength reductions. On the other hand, soils that have been stabilized with lime have some advantages when there is a compaction delay. As opposed to cement stabilized soils, soils stabilized with lime need a curing period so that the lime can diffuse throughout the soil and provide the highest plasticity. Following this curing period, the soils stabilized with lime can be mixed and compacted again allowing for a much higher strength than without these extra steps (Sherwood 1993).

9.5 Moisture Content

Sufficient moisture content is needed for soil stabilization, for both the process of hydration and for effective compaction. Cement, when fully hydrated, can absorb approximately 20% its weight in water from its surroundings (Sherwood 1993). This contrasts with Quicklime (CaO) that can absorb approximately 32% its weight in water from its surroundings (Hebib and Farrell 1999). When moisture is lacking, chemical binders will be in competition with the surrounding soils for moisture. Where the soils have a great affinity for moisture, for example, organic soils, peat, and clay, the hydration of the stabilizing agent may be restricted and this can negatively impact on the ultimate strength of the soil stabilization (White et al. 2005).

9.6 Temperature

The reaction of Pozzolanic is sensitive to changes in temperature. In the field, the temperature changes continuously throughout the day. Pozzolan reactions between binders and soil particles slow down at low temperature and lead to a decrease in the strength of the stabilized mass. In cold regions, it may be advisable to stabilize the soil in the warm season (Sherwood 1993).

9.7 Freeze-Thaw and Dry-Wet Effect

Soils stabilization does not withstand freeze-thaw cycles well. Therefore, at the site, it may be necessary to protect stabilized soils from frost damage. Shrinkage forces in the stabilized soil will depend on the chemical reactions of the binder. Soil, stabilized with cement, is subject to frequent dry-wet cycles due to diurnal temperature changes that can cause stress in the stabilized soil and, therefore, should be protected from such effects (Sherwood 1993) and (Hebib and Farrell 1999).

10   Conclusion

With technological advances and changing economic factors, engineers will have more choices of chemical agents that might be mixed into subgrades as a way of improving the strength, durability and compatibility of the soils. However, performance-based tests must be performed to verify the practicality of any agents used for stabilization. Beyond this, a number of chemicals now in use in the petrochemical industry have not been tested for soil stabilization. Other potential sources of soil stabilization that require research are processes including spray-on techniques and injections that may give rise to practical and economical options. At the same time, the process of global climate change has the potential to adversely affect soil stabilization in terms of both the application and the durability of treatments. As such, it is advisable to review how soil stabilization will be affected by such changes (MacLaren and White 2003). After reviewing the literature, the present research determines that the practical application of the materials discussed here have been proven to increase soil strength or stabilize loose soils. Nonetheless, more research must be conducted to determine the practicality in field conditions, as opposed to concentrating only on experimental research. As ever, the field will benefit from an openness to finding and testing other materials that might be of use for soil stabilization.

References

Addo, J.Q., Sanders, T.G., Chenard, M.: Road dust suppression: effect on unpaved road stabilization (2004)

Al Tabbaa, A., Stegemann, J.A. (eds.): Stabilisation/Solidification Treatment and Remediation: Proceedings of the International Conference on Stabilisation/Solidification Treatment and Remediation. CRC Press, Cambridge, 12–13 April 2005

Baghini, M.S., et al.: Effects on engineering properties of cement-treated road base with slow setting bitumen emulsion. Int. J. Pavement Eng. 18(3), 202–215 (2017)

Bandara, W.W.: The Cement stabilized soil as a road base material for Sri Lankan roads (2015)

Cizer, Ö., et al.: Carbonation reaction of lime hydrate and hydraulic binders at 20 C. In: Proceedings of the First International Conference on Accelerated Carbonation for Environmental and Materials Engineering. The Royal Society (2006)

Cortellazzo, G., Cola, S.: Geotechnical characteristics of two Italian peats stabilized with binders. In: Proceeding of Dry Mix Methods for Deep Soil Stabilization, pp. 93–100 (1999)

Das, B.M.: Chemical and mechanical stabilization. Transportation Research Board. ANSI, B. ASTM D698-Test Methods for Moisture-Density Relations of Soils and Soil-Aggregate Mixtures. Method A (Standard Proctor) (2003)

EuroSoilStab: Development of Design and Construction Methods to Stabilize Soft Organic Soils: Design Guide for soft soil stabilization. CT97-0351, European Commission, Industrial and Materials Technologies Programme (Rite-EuRam III) Bryssel (2002)

Firoozi, A.A., et al.: Fundamentals of soil stabilization. Int. J. Geo-Eng. 8(1), 26 (2017)

Grogan, W.P., Weiss Jr., C.A., Rollings, R.S.: Stabilized Base Courses for Advanced Pavement Design Report 1: Literature Review and Field Performance Data. Army Engineer Waterways Experiment Station Vicksburg Ms Geotechnical Lab (1999)

Hall, M.R., Najim, K.B., Keikhaei Dehdezi, P.: Soil stabilisation and earth construction: materials, properties and techniques. Mod. Earth Build., 222–255 (2012)

Hebib, S., Farrell, E.R.: Some experiences of stabilizing Irish organic soils. In: Proceeding of Dry Mix Methods for Deep Soil Stabilization, pp. 81–84 (1999)

Hicks, R.G.: Alaska soil stabilization design guide (2002)

Jawad, I.T., et al.: Soil stabilization using lime: advantages, disadvantages and proposing a potential alternative. Res. J. Appl. Sci. Eng. Technol. **8**(4), 510–520 (2014)

Kearney, E., Huffman, J.: Full-depth reclamation process. Transp. Res. Rec. J. Transp. Res. Board **1684**, 203–209 (1999)

Keller, I.: Improvement of weak soils by the deep soil mixing method. Keller Bronchure, pp. 23–30 (2011)

Kowalski, T.E., Starry Jr., D.W.: Modern soil stabilization techniques. In: Characterization and Improvement of Soil Materials Session. Annual Conference of the Transportation Association of Canada, Saskatoon, Saskatchewan, Canada (2007)

Lim, N.W.: Thermal stabilization of soils (1983)

MacLaren, D.C., White, M.A.: Cement: its chemistry and properties. J. Chem. Educ. **80**(6), 623 (2003)

O'Flaherty, C.A.: Highways: The Location, Design, Construction and Maintenance of Pavements, Chap. 9, p. 239 (2002)

Okasha, T.M., Abduljauwad, S.N.: Expansive soil in Al-Madinah, Saudi Arabia. Appl. Clay Sci. **7**(4), 271–289 (1992)

Olarewaju, A.J., Balogun, M.O., Akinlolu, S.O.: Suitability of eggshell stabilized lateritic soil as subgrade material for road construction. Electron. J. Geotech. Eng. **16**, 899–908 (2011)

Patel, M.A., Patel, H.S.: A review on effects of stabilizing agents for stabilization of weak soil. Civ. Environ. Res. **2**(6), 1–7 (2012)

Pousette, K., et al.: Peat soil samples stabilised in laboratory–experiences from manufacturing and testing. In: Proceeding of Dry Mix Methods for Deep Stabilization, pp. 85–92 (1999)

Rogers, C.D.F., Glendinning, S.: Modification of Clay Soils Using Lime. Lime Stabilization, pp. 99–112. Thomas Telford, London (1996)

Rogers, J.D., Olshansky, R., Rogers, R.B.: Damage to foundations from expansive soils. Claims People **3**(4), 1–4 (1993)

Sherwood, P.: Soil stabilization with cement and lime (1993)

Umesha, T.S., Dinesh, S.V., Sivapullaiah, P.V.: Control of dispersivity of soil using lime and cement. Int. J. Geol. **3**(1), 8–16 (2009)

White, D.J., et al.: Fly ash soil stabilization for non-uniform subgrade soils (2005)

Sustainability of the Ballasted Track – A Comprehensive Review on Reducing the Use of Mineral Aggregates and the Role of Sub-ballast as a Protective Layer

Cassio E. L. de Paiva and Mauro L. Pereira[(✉)]

Department of Geotechnics and Transportation, University of Campinas
(Unicamp), Campinas, Brazil
celpaiva@fec.unicamp.br, maurolpereira@gmail.com

Abstract. Several aspects must be considered when evaluating the sustainability of a railway, such as the use of mineral aggregates in ballasted tracks. To this day, most of the cargo and passenger railways around the world are built on granular layers of rocky material, which is called ballast. It has the structural function of absorbing and dissipating the vertical stresses resulting from train traffic, at the same time as it allows the quick drainage of rain water and facilitates the correction of geometrical defects of the track's alignment. Nonetheless, ballast grains lose mass due to abrasion, which results in contamination of the layer and differential settlings of the track. Problems also emerge when the railway is constructed without sub-ballast, a blocking and filtering layer located between ballast and subgrade. In developing countries such as Brazil there is still shortage of efficient technology for ballast maintenance, like what is employed in Europe and North America. Likewise, it is common to rebuild or rehabilitate old lines that endure heavy cargo traffic without executing the sub-ballast layer. Considering that ballast is a non-renewable natural material it is compelling that such practices be reviewed and replaced by more sustainable techniques. At the same time, it becomes essential that rocks with adequate properties are used in the construction and maintenance of the railway ballast, so that its life span is the longest possible. The objective of this work is to provoke reflection regarding the rational use of mineral non-renewable resources in the construction and maintenance of rail tracks. A review was made on the desirable properties of rocks to be used as railway ballast. Simultaneously, studies are presented that evaluated the consequences of the absence of sub-ballast in the performance of a railway, as well as different techniques that are used to avoid the interlocking of ballast particles into the subgrade. Finally, recommendations are made to constructors and keepers of rail tracks in developing countries, reinforcing the technical and economic advantages of some already established solutions.

© Springer Nature Switzerland AG 2019
S. El-Badawy and J. Valentin (Eds.): GeoMEast 2018, SUCI, pp. 34–45, 2019.
https://doi.org/10.1007/978-3-030-01911-2_4

1   Introduction

Railway transportation has always been considered a particularly sustainable mode due to its high efficiency in moving great amounts of cargo with low fuel consumption, and therefore low emission of pollutants. This particularity becomes more and more relevant in times of increasing environmental awareness, making railways even more competitive towards the road transportation of cargo and passengers. However, there are other aspects to be evaluated concerning the sustainability of a railway, such as the use of mineral aggregates in ballasted tracks.

Even though railways can be built on different foundations, such as concrete slabs or asphalt, most rail tracks are of the traditional ballasted type, from heavy-haul freight lines to high-speed passenger trains such as the French TGV. One of the main reasons for choosing the granular ballast system is its capacity of absorbing vibrations originated by train traffic, as well as its relatively low construction cost, ease of maintenance works, high hydraulic conductivity of track structure and simplicity in design and construction (Indraratna et al. 2011).

On the other hand, one of the most significant disadvantages is that by being exposed to constant weathering and to repeated loading cycles and particle abrasion, ballast grains lose mass due to friction, which causes the fouling (or contamination) of the layer and differential settlings of the track as a whole. That has been particularly relevant over the past decades, since the traditional rail foundation consisting of one or two granular layers overlying soil subgrade has become increasingly overloaded due to the utilisation of faster and heavier trains, demanding more frequent maintenance cycles (Indraratna et al. 2011). Thus, as traffic demand and the frequency of maintenance cycles increase, so does the use of natural aggregates, which are quarried from non-renewable sources.

It is crucial that environmental concerns also dictate the choice and usage of ballast materials, since the longer they last, the lesser the impact on natural quarries. For that reason, research on the properties of ballast material and the development of construction techniques also contribute to the preservation of natural resources. Several studies have been conducted at the University of Campinas over the last decade approaching the importance and influence of ballast and sub-ballast properties on the performance of the track. Nonetheless, this comprehensive review has the main purpose of looking at those matters from a sustainability point of view, instigating reflection regarding the rational use of mineral non-renewable resources in the construction and maintenance of rail tracks, as well as to shed light on the environmental importance of using the sub-ballast as a blocking layer between the ballast particles and the subgrade.

2   Railways and Sustainability

2.1   Sustainability in the Transport Sector

Awareness of the effects of human activities on the environment and of the shortage of natural resources has led governments and international institutions to the adoption of the term "sustainable development", which was first used by an United Nations Commission that defined it as the kind of economic and social development in which resource use aims to fulfil human needs while preserving the environment, so that future generations can satisfy their needs and enjoy some level of prosperity (United Nations 1987). The United Kingdom government has also launched a strategy for sustainable development that revolves around five key principles: living within environmental limits; ensuring a strong, healthy and just society; achieving a sustainable economy; promoting good governance and using sound science responsibly (UK government 2005).

In the context of civil engineering, according to Fenner et al. (2006), prioritizing sustainable development means that built infrastructure and the provision of associated services should be delivered in such a way that, besides meeting cost effectiveness of the enterprise and excellence and robustness in the application of engineering principles, engineers must also extend their role to ensure that the real needs of all present end users are met, recognising impacts and the opportunity for mitigation and benefit on both the natural environment and future generations.

Governmental and academic institutions worldwide have developed different assessment systems to evaluate the sustainability of transportation infrastructure projects. In Australia, which is one of the global pacesetters regarding heavy-haul railway transportation, the "Transport Environment and Sustainability Policy Framework" was launched (Transport for New South Wales 2013). Its goal is to ensure that their transportation system is sustainable over time and that environmental and sustainability performance is continually improved. That is achieved by monitoring GHG emissions, water, pollution control, noise management, resource management, waste management, material consumption and biodiversity.

A similar assessment system was launched in 2012 by the United States government through the Federal Highway Administration (FHWA), which is called INVEST (Infrastructure Voluntary Evaluation Sustainability Tool) and was developed for voluntary use by transportation agencies to assess and enhance the sustainability of their projects and programs. It is a web-based self-evaluation tool comprised of voluntary sustainability best practices that cover the full lifecycle of transportation services, including system planning, project planning, design, and construction, and continuing through operations and maintenance.

Rail transportation generates one tenth of the carbon dioxide emissions compared to road freight and is also 10 times more fuel efficient; at the same time, it holds the key to improvement of road congestion, since one freight train removes about 150 trucks off the road (Indraratna et al. 2011). Those facts alone are enough to put railway transportation in a strategic position regarding sustainable development of our planet. However, the geotechnical works demanded by the construction and maintenance of rail tracks involve major changes of landform that will persist for centuries, such as rock mining, and the use of large amounts of natural aggregates which carry relatively high carbon impacts (Jefferis 2008). Therefore, it is paramount that such impacts be minimized.

2.2 Environmental Impacts of Aggregate Extraction

Natural aggregates generally are only available on the surface where certain geological processes have occurred (Langer and Glanzman 1993); crushed stone deposits, for instance, are located where particular biogenic-sedimentary, igneous, and metamorphic processes occurred. To produce crushed stone, bedrock is drilled, blasted, and mined from an open-pit quarry or, in some situations, from underground mines. Blasting commonly breaks the rock into pieces that are suitable for crushing.

According to Ratcliffe (1974), the most obvious consequence of mineral exploitation is that of physical change to the land surface, with its cover of soil and vegetation. As mentioned by the author, even when the mines are subterranean, processing and waste disposal almost inevitably produce surface change. Large pits and quarries must be dug in the ground, and in many places these quarries are in production for 25 to 50 years, or even longer. Quarrying creates cliffs, and often screes, and mine shafts or levels become artificial caves. Existing features are removed or obliterated, and there may be significant changes in drainage, as when hollows are excavated or subsidence occurs or when waste tips create new convexities (Ratcliffe 1974). Other problems may occur where finer solid matter from workings enters streams and get redistributed as sediment on banks or alluvial flats at lower levels.

Another study by Langer (2001) on the potential environmental impacts of quarrying carbonate rocks (karst) for aggregate extraction mentioned several consequences, such as the production of dust and noise, disturbance by blasting, damages to habitat, biota and water quality, but it also addressed the cascading impacts. According to the author, in karst environments, aggregate mining may alter sensitive parts of the natural system at or near the site thus creating cascading environmental impacts, which may be initiated by the removal of rock, which alters the natural system, which in turn responds, causing another impact, which causes yet another response by the system, and on and on. It is also mentioned that cascading impacts may be severe and affect areas far beyond the limits of the aggregate operation, manifesting themselves only after mining activities have begun and continuing well after mining has ceased.

As mentioned by the Toronto Environmental Alliance (2008), creating the pits or quarries requires the removal of virtually all natural vegetation, top soil and subsoil to reach the aggregate underneath, which leads to a loss of existing animal wildlife and loss of biodiversity as plants and aquatic habitats are destroyed. Additionally, according to them, pits and quarries disrupt the existing movement of surface water and groundwater, interrupting natural water recharge and can lead to reduced quantity and quality of drinking water for residents and wildlife near or downstream from a quarry site. A study by Winfield and Taylor (2005) showed that less than half of the land disturbed in Ontario for aggregate production between 1992 and 2001 has actually been rehabilitated, meaning that most old pits and quarries were not being properly rehabilitated as required by Canadian law.

Even though industry representatives in developed countries usually state that aggregate quarrying is an activity with relatively low impacts on the environment and society, allegedly because actions are taken to mitigate negative impacts, that is definitely not the case in underdeveloped countries. A thorough study conducted in 2014 in the Nairobi county, in Nigeria, observed that abandoned quarries usually turn into enormous dumping sites for urban garbage, besides being the cause of frequent accidents due to the lack of fences or proper signalling (Eshiwani 2007). On top of that, the surrounding communities are frequently affected by air and sound pollution, and several health problems were observed in nearby residents: statistical tests revealed a direct connection between proximity to the quarries and the occurrence of nasal, eye and respiratory infections, and even malaria.

2.3 Durability of Aggregates in Ballasted Rail Tracks

Among the many functions performed by the ballast, the most important are, according to Selig and Waters (1994): working in a resilient and energy-absorbing manner, reducing vertical pressures from the sleeper contact area to the underlying layers; providing voids among particles to allow the drainage of water and storage of fouling material; resisting lateral, longitudinal and vertical forces, thus keeping the track in its original position; facilitating levelling and lining maintenance operations by allowing tamping of the ballast particles; providing electrical resistance between rails and inhibiting the growth of vegetation.

Service life of the ballast layer, that is, how long the material can endure traffic before it needs to be replaced or replenished, depends greatly on the durability properties of the aggregate, such as hardness and abrasion resistance. According to Raymond and Diyaljee (1979), aggregate degradation is as important, if not more important, than the plastic and resilient deformations as a factor affecting the performance of the conventional track structure.

Physical breakdown depends on the mineralogical composition and internal bonding of the material, as well as weathering. Studies have indicated that the main cause of ballast fouling is the degradation and breakage of its larger and once uniformly-graded particles when they are subject to repeated loadings due to the passing of trains (Selig et al. 1988, 1992). Currently, railroads use three abrasion tests: the Los Angeles abrasion (LAA), Mill Abrasion and the Deval attrition tests (Selig and Waters 1994). The Sulphate Soundness test is also used to examine the resistance to chemical action of Sodium Sulphate and Magnesium Sulphate (salt), and that index is used to predict ballast reaction to weathering. Table 1 illustrates three international railway specifications and their requirements regarding properties related to durability. Those standards are from Brazil (ABNT 2011), USA (AREMA 2009) and Australia (Standards Australia 1996).

Table 1. Specified values for ballast properties according to international standards

Standard	LAA (%)	Sulphate soundness (%)	Specific gravity (kg/m^3)	Water absorption (%)	Non-cubic particles (%)
Brazil	≤ 30.0	≤ 10.0	≥ 2500	≤ 0.8	≤ 15
USA	≤ 35.0	≤ 5.0	≥ 2600	≤ 1.0	≤ 5
Australia	≤ 25.0	-	≥ 2500	-	<30

The LAA test is particularly adequate for measuring tenacity, which is the ballast's resistance to breakage of its particles into smaller pieces due to impact and abrasion. It is largely used when a general overview of the aggregate's properties is required, and it can be easily performed. Gaskin and Raymond (1976) found that the LAA index showed great correlation with other hardness and tenacity tests and therefore the authors suggested that it could be used as a sole test to assess ballast properties. Other studies also indicate that the LAA test correlates well with the abrasion caused by tamping maintenance operations (McDowell et al. 2005; Aursudkij 2007).

Several international standards for railway materials consider the Los Angeles abrasion index as an important factor regarding the durability and service life of ballast. Specifications in Brazil, Australia and the United States demand that the LAA index must be lower than 30%, 35% and 25% respectively, or else that aggregate is not adequate (ABNT 2011; AREMA 2009; Standards Australia 1996). Therefore, the LAA index may be considered an important measurement when estimating the service life of the ballast layer and therefore its sustainability.

3 Importance of the Sub-ballast

3.1 Functions and Characteristics

Sub-ballast is the name given to the layer of granular material placed on top of the subgrade and under the ballast, and it is usually comprised of well-graded crushed rock or a mixture of gravel and sand. According to Selig and Waters (1994) it is responsible for preventing the penetration of larger ballast grains into the subgrade soil, thus acting as a blocking layer, as well as stopping the upward migration of subgrade particles (fines) into the ballast, therefore acting also as a filter. Other functions include extending the subgrade frost protection and permitting drainage of water that might flow upward from the subgrade. Profillidis (2014) adds that the sub-ballast also has the role of further distributing the vertical stress that are transmitted from the ballast layer above down to the subgrade over a wider area, as well as allowing the quick runoff of rainwater by granting a transverse slope of 3 to 5% to the upper surface of the subgrade.

Indraratna et al. (2011) reminds that in current design approaches, the main role of sub-ballast is to protect a soft (low quality) subgrade soil, such as a compressive clay, from being excessively loaded, and for that reason the sub-ballast layer is compacted to a much higher stiffness than the natural soil formation, such that the load distribution to the underlying subgrade is significantly reduced as well as being uniform. However, according to the same authors, another key role of the sub-ballast is to act as a drainage layer and an effective filtration medium since drainage plays a significant role in the stability and safety of a track substructure. Therefore, the sub-ballast hydraulic conductivity should be at least an order of magnitude smaller than that of the ballast, while in order to drain water seeping from the subgrade, the sub-ballast should also have a hydraulic conductivity greater than the subgrade.

3.2   Absence of Sub-ballast and Its Consequences to Track Performance

As presented above, the sub-ballast layer is critical for the appropriate performance of the rail track. Unfortunately, in Brazil it is still common to rebuild or rehabilitate old lines that are subject to heavy traffic without executing the sub-ballast layer. The consequence, like explained by Selig and Waters (1994), is that ballast placed directly in contact with the fine-grained soil or soft rock subgrade, when in the presence of water, will wear away the subgrade surface and form a slurry. The slurry will then squeeze upward through the ballast voids which inevitably causes ballast contamination and sinking of the superstructure. Usually the solution given in Brazil to both a ballast layer that reaches the end of its life span and to the caved-in track is simply heightening the track, by adding new gravel on top of the fouled ballast, then lifting the track with tamping machines, as shown in Fig. 1. That practice not only does nothing to permanently solve the problem but also contributes to the unrestrained and unsustainable usage of mineral aggregates. Additionally, it leaves the track bed with no ballast shoulder, making it much more prone to buckling and alignment defects.

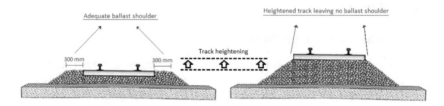

Fig. 1. Track heightening and consequent loss of shoulder width

A series of studies were conducted at the University of Campinas, in south-eastern Brazil, in order to evaluate the influence of sub-ballast in the track's performance and service life. Paiva et al. (2011) ran a comparative analysis between two sections of the same railway, where only one of them had been built with a sub-ballast layer. They were both subject to identical loading states by passing a locomotive repeatedly at constant speeds, and conditions of the ballast was evaluated as well as vertical strains, measured by a laser deflectometer. As presented by Table 2, results showed that, even though the maximum deflection was similar for both conditions, the deflection basin of the line without sub-ballast (Line A) presented much higher strains, and its ballast was highly contaminated with fine particles, while the other line (Line B) had clean ballast.

Rodrigues (2016) simulated a track without sub-ballast by applying dynamic and static loadings to ballast samples and evaluated the penetration of those particles into the subgrade. Dry and water-saturated samples were subject to a vertical stress of 412 kN/m^2, equivalent to that on the sleeper-ballast surface, with dynamic load tests of 10,000 cycles. In samples with the least compacted subgrade (80% of the Proctor Energy) ballast particles penetrated up to 13 mm under static loading and 47 mm under dynamic loading. Even samples with the highest subgrade compaction (98% of the Proctor Energy) showed penetration of up to 10.2 mm, as shown in Table 3.

Table 2. Strain measurements (Paiva et al. 2011)

Distance between the points of study and measurement (cm)	Deflection measurements (mm)	
	Line A	Line B
0	4.41	4.34
60	4.41	4.30
120	4.40	4.05
140	4.35	3.95
180	4.25	3.45
210	4.18	3.06
240	4.11	2.67
300	4.07	1.85
260	3.80	1.16
B – Deflection basin length (m)	5.31	4.85
Ballast Coefficient - c (N/mm^3)	0.0313	0.036

Table 3. Ballast interpenetration into the subgrade due to dynamic loading, under different conditions (Rodrigues 2016).

	Sample number					
	1	2	3	4	5	6
Sample condition	Dry	Soaked	Dry	Soaked	Dry	Soaked
Sample density (KN/m^3)	12.7	12.7	14.3	14.3	15.7	15.7
Compaction (%)	79.2	79.6	89.3	89.6	98.1	97.9
Interpenetration (mm)	20.17	47.15	10.8	16.5	9.04	10.21

It is important to notice, however, that even with the placement of high quality ballast and sub-ballast layers, instability or failure of the subgrade will inevitably result in unacceptable distortion of track geometry and alignment (Indraratna et al. 2011). That means that a subgrade with low bearing capacity and high resilience will provoke displacements on the entire track, no matter how thick or compacted the sub-ballast is. Therefore, proper design and sizing of all layers of the substructure is imperative to the railway's adequate performance and service life.

3.3 Alternatives to the Granular Layer

Given the importance of having a stiff subgrade for the proper performance of the rail track, one broadly explored alternative to the granular sub-ballast is the bituminous substitute. Teixeira et al. (2009) consider it a structural improvement to the ballasted track since it reduces its maintenance needs without increasing too much of its construction costs and it keeps the subgrade's moisture content closer to optimum during all year. Those authors evaluated the impact of bituminous sub-ballast in improving conventional high-speed ballasted tracks in Spain. Their results show that this solution might bring relevant advantages in terms of subgrade protection and life cycle, track differential settlement and track dynamic performance. At the same time, they affirm that the cost of the bituminous sub-ballast in certain parts of Spain is similar to the cost of the granular solution. According to Selig and Waters (1994), there are several advantages to replacing the granular sub-ballast layer by asphalt concrete: the stress reduction function is maintained; interpenetration of ballast into the subgrade is eliminated, as well as upward migration of fines; water shedding is maximized. Also, since the ballast layer is kept, track geometry can be adjusted by tamping the ballast.

Alternatively, granular sub-ballast can also be replaced by geotextile materials and geogrids. Horníček et al. (2010) studied the effects of three types of reinforcing geosynthetics (woven, welded, and extruded geogrids) inserted directly under the ballast, comparing the values of settlement and bearing capacity of individual layers with those of a nonreinforced construction. Tests were performed in laboratory as well as on a trial section of a single-rail track with a length of 200 m, well known for persisting long-term problems with track geometry. Laboratory experiments performed on a track bed model in a 1:1 scale showed an increase in the bearing capacity of the track bed, both under short-term static loading as well as long-term cyclic loading. On the trial section, partial results after one year of monitoring have also been positive, with control car inspections showing smaller and fewer imperfections in the geometric parameters.

Another low-cost option for increasing the stiffness of the subgrade layer is treating the soil with chemical additives. That type of solution has been studied for several years, and it was approached in a very detailed report published by Brazilian researchers under the Committee on Tropical Soils of the ISSMFE (1985). In that report, which addressed the peculiarities of tropical soils used as construction materials in several fields, they already mentioned the advantages of using lime to correct soil acidity and plastic behaviour, especially when clay particles are present. It was also said that adding the proper percentage of Portland cement can drastically modify the properties of soils mixtures, improving their mechanical performance.

A study by Joaquim (2017) treated two types of tropical soils, that had very poor mechanical performance, with different levels of cement and lime. In that research, soils were categorized according to MCT (Miniature, Compacted, Tropical), a classi-fication system developed in Brazil by Nogami and Villibor (1981) that is considered more accurate on classifying tropical soils than the existing conventional systems, due to those soils' very particular properties. Paige-Green et al. (2015) considered MCT the specification that has been the most successfully applied to the greatest length of roads in the world, regarding lateritic soils. Results found by Joaquim (2017) showed that the addition of 4% of cement or 8% of lime to lateritic, clayey soils (called LG′ in the MCT system) and non-lateritic-clayey (called NG′ in MCT), compacted in the normal and intermediate energies, respectively, made those soils meet the Brazilian specifications for unpaved roads, which are similar to the railway subgrade. Table 4 shows that the mass loss due to immersion in water was completely stopped for both types of soils, using either of the additives.

Table 4. Mass loss by immersion in water for different soils and different additive contents (Joaquim 2017)

MCT classification	Condition	Mass loss	Compaction energy
LG'	Natural Soil	37.1%	5.6 MN/m^2
	4% of Cement	0.0%	5.6 MN/m^2
	8% of Lime	0.0%	5.6 MN/m^2
NG'	Natural Soil	49.3%	5.6 MN/m^2
	Natural Soil	0.7%	12.4 MN/m^2
	4% of Cement	0.0%	12.4 MN/m^2
	8% of Lime	0.0%	12.4 MN/m^2

4   Conclusions

Through the information gathered in this research it is possible to conclude that the choice of ballast material and the construction and proper maintenance of a sub-ballast layer are two key factors regarding the sustainability of a railway project, increasing its durability and improving its performance. Using aggregates with the highest possible resistance to abrasion, which is a property easily and reliably evaluated by the Los Angeles test, extends the service life of the track's substructure, thus minimizing the extraction and usage of mineral non-renewable resources. That practice has positive consequences on conservation of the environment as well as on the health and well-being of communities that surround pits and quarries, not to mention the economic gains of railways and contractors.

Similarly, it must be ensured that new rail projects are constructed with an adequate sub-ballast layer with separation and filtering properties. Aside from preventing a series of other problems to the track's performance, it ensures that ballast particles will not penetrate into the subgrade, thus provoking ballast contamination and demanding its replacement or the addition of large amounts of new aggregate. There are several options available today, ranging from the simplest chemical treatment of the soil with lime to asphalt layers or high-tech geogrids and geotextiles, as well as the traditional granular sub-ballast.

Therefore, it can be said that one way of achieving sustainability of railway projects is guaranteeing that tracks will have longer service lives, with less frequent maintenance cycles and by diminishing the need for addition of new ballast material.

References

American Railway Engineering and Maintenance-of-Way Association (AREMA): Manual for Railway Engineering (MRE), Chap. 1: Roadway and Ballast, Maryland, USA (2009)

Aursudkij, B.: A Laboratory Study of Railway Ballast Behaviour under Traffic Loading and Tamping Maintenance. Doctoral Thesis – University of Nottingham, Nottingham, England (2007)

ABNT NBR 5564. Via férrea – Lastro ferroviário – Requisitos e métodos de ensaio, Rio de Janeiro (2011)

Committee on Tropical Soils of the ISSMFE. Peculiarities of Tropical Soils Used as Construction Materials. Topic 4.2: Roads. Progress Report 1981–1985. First International Conference on Geomechanics in Tropical Lateritic and Saprolitic Soils (1985)

Eshiwani, F.: Effects of quarrying activities on the environment in Nairobi county: a case study of Embakasi district. Master Thesis, University of Nairobi (2007)

Fenner, R.A., Ainger, C.M., Cruickshank, H.J., Guthrie, P.M.: Widening engineering horizons: addressing the complexity of sustainable development. Eng. Sustain. **159**(4), 145–154 (2006)

Gaskin, P.N., Raymond, G.P.: Contribution to selection of railroad ballast. J. Transp. Eng. Div. **102**(TE2), 377–394 (1976)

Indraratna, B., Salim, W., Rujikiatkamjorn, C.: Advanced Rail Geotechnology – Ballasted Track. CRC Press/Balkema, Leiden (2011)

Jefferis, S.A.: Moving towards sustainability in geotechnical engineering. In: Proceedings of Geocongress 2008, American Society of Civil Engineers, New Orleans, United States, pp. 844–851 (2008)

Hornícek, L., Tyc, P., Lidmila, M., Krejciríková, H., Jasanský, P., Brešt'ovský, P.: An investigation of the effect of under-ballast reinforcing geogrids in laboratory and operating conditions. Proc. Inst. Mech. Eng. Part F J. Rail Rapid Transit **224**, 269–277 (2010)

Joaquim, A.G.: Study of two tropical soils improved with cement or lime for employment in upper layers of unpaved roads. Master's Thesis, School of Civil Engineering, Campinas State University, Brazil (2017)

Langer, W.H.: Potential Environmental Impacts of Quarrying Stone in Karst – A Literature Review. Open-File Report OF-01-0484, U.S. Geological Survey (2001)

Langer, W.H., Glanzman, V.M.: Natural aggregate – building America's future. U.S. Geological Survey Circular, n. 1110, Denver, United States (1993)

Mcdowell, G., Lim, W., Collop, A., Armitage, R., Thom, N.: Laboratory simulation of train loading and tamping on ballast. Proc. Inst. Civ. Eng. Transp. **158**(2), 89–95 (2005)

Nogami, J.S., Villibor, D.F.: Uma nova classificação de solos para finalidades rodoviárias. In: Proceedings of the Simpósio Brasileiro De Solos Tropicais Em Engenharia, Rio de Janeiro, pp. 30–41 (1981)

Paige-Green, P., Pinard, M., Netterberg, F.: A review of specifications for lateritic materials for low volume roads. Transp. Geotech. **5**, 86–98 (2015)

Paiva, C.E.L., Peixoto, C.F., Correia, L.F.M., Aguiar, P.R.: Evaluation between two Brazilian railway tracks. J. Civ. Eng. Archit. **5**(9), 772–778 (2011)

Profillidis, V.A.: Railway Management and Engineering, 4th edn. Ashgate-Publishing Group, Surrey (2014)

Ratcliffe, D.A.: Ecological effects of mineral exploitation in the United Kingdom and their significance to nature conservation. Proc. R. Soc. **339**, 355–372 (1974)

Raymond, G.P., Diyaljee, V.: Railroad ballast load ranking classification. J. Geotech. Eng. Div. **105**(10), 1133–1153 (1979)

Rodrigues, M.: Colmatação de Camadas Granulares por Finos do Subleito em Condições Deficientes de Compactação e Drenagem. Bachelor's Thesis, School of Civil Engineering, Campinas State University, Brazil (2016)

Selig, E.T., Dellorusso, V., Laine, K.J.: Sources and Causes of Ballast Fouling. Report R-805. Association of American Railroads, Chicago, United States (1992)

Selig, E.T., Collingwood, B.I., Field, S.W.: Causes of Fouling in Track. AREA Bull. **717**, 381–398 (1988)

Selig, E.T., Waters, J.M.: Track Geotechnology and Substructure Management. Thomas Telford Publications, London (1994)

Standards Australia. AS 2758.7: Aggregates and rock for engineering purposes, Part 7: Railway ballast, NSW, Australia (1996)

Teixeira, P.F., Ferreira, P.A., López Pita, A., Casas, C., Bachiller, A.: The use of bituminous subballast on future high-speed lines in Spain: structural design and economical impact. Int. J. Railw. **2**(1), 1–7 (2009)

Transport for New South Wales: Transport Environment and Sustainability Policy Framework. New South Wales, Australia (2013)

Toronto Environmental Alliance: The Environmental Impacts of Aggregate Extraction (2008). http://www.torontoenvironment.org/gravel/impacts

UK Government: Securing the Future - UK Government Sustainable Development Strategy. The Stationary Office, London (2005)

United Nations General Assembly: Report of the World Commission on Environment and Development: Our Common Future, Oxford (1987)

Winfield, M.S., Taylor, A.: Rebalancing the load: the need for an aggregates conservation strategy for Ontario, pp. 8–9. The Pembina Institute (2005)

Impact of Climate Change on Traffic Controls in Developing Countries

Bongeka L. Mbutuma[(✉)], M. Mostafa Hassan, and D. K. Das

Sustainable Urban Roads and Transportation Research Group, Department of
Civil Engineering, Central University of Technology, Bloemfontein, Free State,
South Africa
mbutumabl@gmail.com, {mmostafa,ddas}@cut.ac.za

Abstract. Severity of climate condition is becoming more frequent, therefore it
much important to be very clear about the influence of climate on road users
daily travelling pattern. Climate change has much effect on the aspects of
transportation systems such as. Climate change is three dimensional which are
exposed, sensitive and low adaptive capacity to traffic highway. Although these
dimensions of climate impact on traffic highway are measurable and predomi-
nant, they need potential adapters to monitor and support the adaptive response,
potential persisted to monitor population trend and high latent risk to monitor
the environment. This includes traffic demand, traffic safety and traffic operation
relationships. Furthermore, research as established that severe climate change
such as winter storm brings about 25 times the normality of vehicular crashes,
this is however, much higher than the risk brought by the road users behavior.
The risk of drivers involved in a vehicular crashes becomes more intense as a
result of extreme weather conditions. Thus, there is a need for the transportation
related authorities to manage and restrict the use of highways during extreme
condition in order to vehicle operation cost and other related cost. Nevertheless,
the first step by transportation agencies to manage transportation systems to
minimize the weather impact is to quantify its impact on traffic on highways.

Thus, this paper presents a review of the impact of climate condition on traffic
demand, traffic safety and traffic operation relationships. Using a case study area
of Eastern Cape Province in South Africa, accident reports were collected from
secondary sources and analyzed. Overall, a relationship is drawn on the impact
of climate condition and travel fatality. Furthermore, traffic regulation recommendations based on the impact of climate conditions will be drawn.

Keywords: Climate change · Traffic fatality · Traffic controls

1 Introduction

Climate change is a major global drastically problem that is likely to worsen and if
temperature and rainfall continue to rise at current rates, climate change threatens the
human and infrastructure development. Poor communities are already the most vulnerable to harsh and variable climates such as extremes of rainfall and temperature,
shifts in growing seasons and unpredictable. Also violent and sudden climatic events
are going to increase with temperature (Field et al. 2014). As the global warming levels

© Springer Nature Switzerland AG 2019
S. El-Badawy and J. Valentin (Eds.): GeoMEast 2018, SUCI, pp. 46–57, 2019.
https://doi.org/10.1007/978-3-030-01911-2_5

increase, heavy rainfall, heat waves and drought will become more pronounced. South African is currently facing the risk of more severe and protracted droughts and periods of extremely high and extreme low rainfall (Field et al. 2014). Secondary sources such as climate models are broadly consistent in predicting that rainfalls will be heavier, specifically in the wetter areas of tropical Africa, which will then increases the flood hazards.

The regular road users understand that climate change matters to traffic facilities and traffic controls. Anytime during the snowy mornings or nights, the radio and television provide commuters with copious reports regarding the amount and intensity of wind speeds and heavy rainfall in very dangerous areas. It is known that television regularly shows live videos of vehicular crashes and congested freeways in some areas as commuter slip and slide their way to various destinations through the snow (Maze et al. 2006). Through such experience, commuters should have a better understanding that a forecast of wind speed and snow accumulation is likely to imply a very negative impact for travelling purposes. This kind of inclement weather results in most hazardous trips and travel times are longer and less reliable. It then gives a driver no choice but to adjust their travelling schedule to take delays and unreliable travel times into account. Most travellers are required to cancel or differ trip (Maze et al. 2006).

1.1 Background

South Africa is already a climatically sensitive and stress water country. The country is arid or semi-arid and the whole country is subjected to floods and drought, thus the variation in the rainfall or temperature is exacerbated to stress environmental factors. The impact of climate change will thus worsen the serious lack of surface and ground water resources that may later alter the magnitude, timing and distribution of storms that produce floods. Floods may be a huge disturbance to traffic controls and may lead to high traffic fatality (Field et al. 2014).

South African government is then required to have a moral obligation to respond to the impacts and to take a lead to develop both mitigation and adaptation strategies. The implemented strategies must not only focus on the ecological and economical aspects but most importantly on social, environmental and sustainable aspects of climate change that will affect road infrastructure and traffic controls. For about seventy percent of the land in South Africa consists of natural and semi-natural ecosystems which proved rangelands for substantial herbivore species. The modelling formation suggests a general arid type of land, mostly in rangelands that are already marginal. In such situation the tree encroachment into the grassland areas is likely to increase due to the elevated CO_2 concentrations and the increase in temperature (Field et al. 2014). The frequency of fire outbreaks are predicted to increase significantly, therefore roads users are forced to cancel or delay their road trips. It is likely of the road infrastructure or surface deterioration to occur.

The inclement weather such as snow, fog high wind, extreme cold and heavy rainfall have an intensive impact on traffic controls during crucial proportion of the year, thus crucially impacts annual safety performance and the annual vehicle capacity of urban highways (Takeuchi 1978). According to the National Climate Change Strategy (NCCS) used the Global Climate Model, South Africa Climate Change has been predicted to a continental warming of between 1 and 3 degrees Celsius with a broad reduction of approximately 5 to 10% of current rainfall (Mallick 2015). Although with a higher rainfall in the east and drier conditions in the west of South Africa. There is an abnormal extension of the summer season characteristics which are mainly caused by the climate change. Also increased summer rainfall in the northeast and southwest, but have a reduction of duration of summer rains in the northwest and an overall reduction of rainfall in the southwest of South Africa During the winter season the country has a nominal increase in rainfall in the northeast and increased daily maximum temperature in summer and autumn in the western part of the country (Mallick 2015).

1.2   Socio-economic, Environmental and Health Factors as Traffic Controls

In transport sector, vehicles rely on oil with about 94% of transport fuel being petroleum products and that makes it a key area of energy security concern. Oil dominants as a major source of harmful emissions that affects air quality that have a huge effect on climate change. In most developing countries, a combination of public transit and cycling infrastructures land-use measures and pricing are projected to lead to notable co-benefits such as precise hierarchical structure with directive features to assist firms in identifying the key practices that will result in firm achievement (Griffiths and Rao 2009). Previous studies have addressed co-benefits as they relate to climate change, reducing energy consumption, preventing disadvantageous environmental impact, improving public health, developing relevant policies and considering the practice of co-benefits in terms of a firm operation (Wu et al. 2018).

Several studies have addressed co-benefits as they relate to climate change, preventing adverse environmental impacts, reducing energy consumption, improving public health and developing relevant policies (Woodcock et.al. 2009; Kwon et al. 2016; Liu 2017), but their theoretical linkage to TBL remains in the infancy stage. In addition, few prior studies considered the practices of co-benefits in terms of a firm's operation.

Then also the improvement of energy security, reduction of fuel spending, fewer road accidents, less congestion, less air pollution and noise-related stress, and increasing public health from more physical activities. Only few studies did the assessment on the associated welfare effects comprehensively and are delayed by data uncertainties. More fundamental in the epistemological uncertainty attributed to various social costs that result in the large range of plausible social cost and benefits. The improvement of mobility access, traffic congestion, traffic safety, traffic demands and energy security are extremely important policy objectives that can possibly be influenced by mitigation actions and adaptions (Jacobsen 2003; Hultkrantz et al. 2006).

1.2.1   Traffic Congestion

Traffic congestion is an important aspect for decision makers and the authorities, particularly in the local level and as it negatively affects journey times, speed and substantial economic cost (Goodwin et al. 2004; Duranton et al. 2011). The assessment of social cost in public health is recognized as at contested area when it is furnished as disability-adjusted life years (DALYs). It is then confirmed that in some regions such as London in the United Kingdon and Delhi in India, a reduction in CO2 emissions through an increase in active travel and less use of Internal Combustion energy (ICE) vehicles has given an associated health benefit (Woodcock et al. 2009). It is consistently found across the studies that public health benefits can be induced by modal shift from Light Duty Vehicles (LDVs) to non-motorized transport or public transport and also improves the air quality. Modal shift can reduce traffic congestion, therefore can simultaneously reduce Greenhouse Gas (GHG) emissions and short-lived climate forcers. Some action tends to seek to reduce traffic congestion can induce additional travel demand such as the construction of new roads to increase capacity (Goodwin et al. 2004).

1.2.2 Traffic Safety

Globally the increase in motorized road traffic in most countries have an increasing incidence of accidents with the estimate of about 1.27 million people killed yearly and statistically, the amount of 91% occur in middle and low income countries (TFTAHC 2000). Furthermore, 20 to 50 million people are suffering from severe injuries due to road accident. It is estimated that by the year 2030, the road traffic injuries will constitute the fifth biggest reason for premature deaths.it is argued to increase the efficiency of the vehicle fleet, so that it can also positively affect the crash-worthiness of vehicles in if more stringent safety standards are adopted along with the improved efficiency standards (TFTAHC 2000). In most developing countries the lack of access to safe walking, cycling and public infrastructure remains an important element affecting the success of modal shift strategies (Tiwari and Jain 2012).

1.2.3 Access and Mobility

Multimodality is likely to foster improved access to transport services especially for most vulnerable members of the society the improved mobility usually provides access to jobs, markets, school, hospitals and other facilities (Liang et al. 1998). The efficiency transport and modal choices do not only increase access and mobility but also positively affect transport costs for certain individuals and businesses. It also creates the affordable and accessible transport system foster productivity and social inclusion.

1.3 Climate Change Impact on Traffic Controls and Facilities

The literature mostly describes the estimated impact of rain, heavy wind, extreme cold, extreme heats and reduced visibility, the less of snow fall.

There are three predominate traffic variables that climate impacts:

- **Traffic demand**: The challenge if inclement of climate some trips will be defeated, eliminated or postponed.
- **Traffic Safety**: Vehicular crash rate increases drastically during inclement climate.
- **Traffic operations**: Changes in the relationship between traffic speed, volume density resulting in reductions of capacities.

1.3.1   Impact of Climate Change on Traffic Demand

Previous researchers have found that traffic volume declines during heavy rainfall and severe wind speed. Such activity may be possible due to several reasons, likely to be motorists diverting, postponing or cancelling trips before after trips. Traffic volume reduces due to have rainfall or winter storm across varied intensity of snow fall, or iced roadways during time of the day, days of the week or consecutive weeks (Liang et al. 1998). The traffic reductions is ranged between 7% and 60% depending on the category of weather event. This reduction is dependent to the increase with the total volume of snowfall and rainfall. The reduction of volume becomes smaller during peak travel period and weekdays than off-peak period and weekends.

There are certain studies carried out in different countries, those in tropical areas versus temperate areas, developed versus developing countries that may also yield distinct results since inhabitants in various region have different climate mitigation and adaption strategies resulting in perceptions of cold, warm or normal weather conditions (Liu 2017). The variation in land use, culture and transportation network may also affect the impact of climate change in various region. Most studies have focused on a single travel behavior dimension, such as trip distance or mode choice trip and few studies focused on developed models involving several travel behavior dimensions (Liu 2017). Travelling is categorized as the derived demand of activity participation and studies have jointly model activity time use and activity participation with travel mode choice and trip distance that sustainable describe climate impact on travel demand (Liu 2017). The quantification impact of climate change on travel behaviors have been prioritized and the resulting information is very rarely integrated into the existing transportation planning process. By the given significant climate impacts on travel behavior, it is likely that future traffic demand could vary in a warmer climate scenario and thus lead to variations in traffic flow, transit ridership road congestion emission and safety (Böcker et al. 2013). Figure 1 presents the dynamic interactions between Travel behavior and climate.

1.3.2   Impact of Climate Change on Traffic Safety

Impact of climate change events on traffic safety are varied and the is an indication that vehicle crash rate increases as roadways become iced, wet, during a heavy storm or extreme hot heatwaves (Brodsky and Hakkert 1998; Golob and Recker 2003). Road accident fatality and severe injury rate on wet or iced roads cab be several times greater than on the normal roadway conditions. It is globally estimated that the fatality and injuries have been increased by approximately 25% during extreme weather conditions (Kwon et al. 2016). But, in some cases, vehicular crashes become less severe than their counter part during clear weather conditions. This reduction in the relative rate of severe and fatal crashes are probably due to vehicles reducing their speed due to drastically storm, wind speed or heatwaves, which reduces the severity of road crashes. In the event stormy weather conditions, multiple vehicle crashes are being resulted by lane changing and merging movements (Golob and Recker 2003).

For example, South Africa consist the majority of unpaved roads that are very much affected by climate change. The road and bridges are underdeveloped and there is a lack of maintenance. The past few years, the is a challenge of bridges over floating water during heavy storm, that leads to people of the communities fail to reach their home or travel to meet their needs. According to the South African Police Service

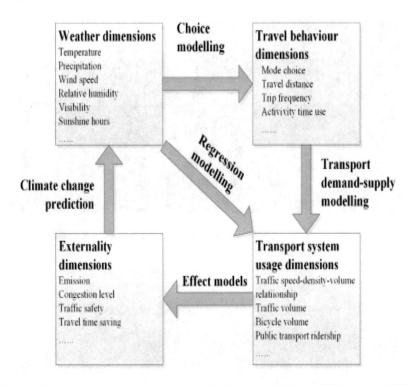

Fig. 1. Dynamic interactions between travel behavior and climate (Böcker et al. 2013).

(SAPS) report, every time during the storm event 1 out 100 000 people in South Africa are drowning trying to pass the bridge. Vehicles are being stuck on the mud due to unstable soil on the road. The fatality and severity rate is drastically increasing due to vehicles swinging and overturn. One of the incidents. Figures 2 and 3 presents the current scenario on effect of climate change on unpaved roads and low bridges.

Summarizing the information obtained from South African Police Services (SAPS) data base. The selected data was collected between November 2017 and January 2018 (12 weeks report) by SAPS, and only recorded the major and fatal accidents. Normally, during non-storm weather events, the crash rate is approximately 0.054 crashes per million vehicle kilometer. The data across, the 57 extreme kinds of weather, states that the storm crash rate of 0.591 crashes per million vehicles kilometers on rural unpaved roads, that indicates the crash rate increases by 11 times during extreme weather events. In this paper Poisson distribution Model have been used, where the dependent variable is the probability of the observed number of crashes and the independent variable are the characteristics of weather such as heavy rain, extreme, very hot heatwaves, wind and snowfall. The results state that the climate change intensity have a positive and statistically significant relationship to the number of crashes (Marchini 2008).

Fig. 2. Low bridges over flowing water due to heavy rain in Eastern Cape, South Africa (Taken by authors).

Fig. 3. Vehicles stuck on the muddy roads in Eastern Cape, South Africa (Taken by authors).

$$P(X) = N - \frac{e^{-m}m^r}{r!} \qquad (1)$$

Where:

P (x) = Number of drivers having accident r accidents
M = average rate of occurrence of event
E = Base of logarithm
N= number of drivers

Calculation
Summarized data:

N = 5 is the average number of historical accidents per week.
: r =
: p = 0.0067
: M = 715 km
: m = PM = 0.0067x747 = 5.0049

$$P(x) = N - \frac{e^{-m}m^r}{r!}$$
$$= 1 - P(x = 3)$$
$$P(x) = 1 - \frac{e^{-5}5^0}{0!} + \frac{e^{-5}x5^1}{1} + \frac{e^{-5}x5^2}{2} + \frac{e^{-5}x5^3}{3}$$
$$= 1 - (0.0067(1 + 5 + 512.5 + 41.67))$$
$$= 0.591$$

More flooding and water logging are being experienced due to frequent increase of extreme rainfall events. Coastal cities such as Durban, East London, Port Alfred, Port Elizabeth and Cape Town will be more affected by more flooding during extreme rainfalls events due to sea level rise in the near future. It can be remarked that most imminent impact of climate change on traffic controls in South Africa will be due to increase temperature, rainfall and snowfall extremes (Keay and Simmonds 2006).

1.3.3   Impact of Climate Change on Traffic Operations

Traffic operations are the functions of traffic density relative to vehicle per lane per kilometer, traffic speed measured by kilometer per hour, road markings and traffic control devices. Traffic Flow Theory and Highway Capacity 2000 defines freeway capacity as the maximum flow rate that can be expected to be achieved repetitively a single freeway location, similar roadway location, traffic control condition without breakdown. The maximum flow capacity of a freeway segment is dependent on the speed of the traffic stream and the inverse density. During inclement conditions, the weather impacts the capacity of the freeway segment that leads the road users to moderate their driving speed and increase the headways between vehicles (Jaroszweski et al. 2014).

Several researcher have measured the extent to which the road capacity and traffic operation are impacted by climate change. The main focus are the parameters such as speed, highway and low volume roads capacity and detector occupancy related to traffic flow during drastically weather conditions (TFTAHC 2000). It is important to measure capacity reductions during wet or snowy road conditions as a result of change in traffic flow characteristics such as lower density and speed when urban facilities are operating near or above capacity (Brilon and Ponzlet 1996). Although, main focus is speed reduction only, because measurement were taken from the uncongested highway segment or rural highway segment (Liang et al. 1998).

1.4   Implementation of Transport Sector Mitigation and Strategy Due to Climate Change

The remarkable mitigation benefits can be instituted in the transport sector, given the alignment intentions for controlling climate change emissions and those related strictly to transportation operations and traffic controls improvements. In the year 2000 the transportation sector was accountable of about 19 percent of South Africa greenhouse gas emission, also obtained the second most significant source of greenhouse gas emission in the middle and high income countries (Shaw et al. 2014).

In South Africa, the energy intensity is particularly high due to the extensive use of synthetic fuels and their production process. Measures of response should then include addressing issues of urban and peri-urban planning in relations to both passengers and commercial transport. Firstly, the passenger transport side, the significant challenge is to improve the currently inadequate public transport system significantly enough to retain 86% of daily commuters it currently carries (Shaw et al. 2014). The number of initiative challenges was being addressed by the National Department of Transport (NDOT). The emissions from the commercial side such as freight transport, further research is required into the range of policy alternatives for influencing the modes of transport and their practical and economic implications for South Africa. The mitigation alternatives comprise energy efficiency improvement, fuel switching, public transport initiatives and new propulsion technologies as shown on Fig. 4.

Coordinating planning and investment in the road infrastructure and services that take accountability of climate change and other environmental issues provide South Africans access to reliable and safe travelling, healthy living, secure road infrastructure, clean water and decent sanitation and making communities resilient to the impact of climate change and less socio-economically vulnerable. South Africa have implemented the reduction of its carbon emission, in line with its international commitments. While maintaining its competitiveness in the global economy by cautiously managing investments in local and regional renewable energy resources and aggressively condoning just equitable trading arrangement (Shaw et al. 2014).

2   Discussion

It is definitely acquired by the study that climate change significantly influences transportation system in terms of both supply and demand of traffic. The transportation policies need to aim at the construction of socially, economically, and environmentally sustainable transport system and infrastructure. The understanding the impact of climate change on individual travel is necessary, therefore active transport modes such as cycling, walking or high capacity occupation transits may be promoted (Öörni and Kulmala 2012).

The fact that future climate will become more intense and extreme, with the understanding on how climate condition affects the traffic controls is vital for planners and decision-makers to achieve a sustainable transportation system to accommodate the future climate change. This paper has assessed and summarized this contribution from the individual behavioral condition and despite extensive empirical evidence

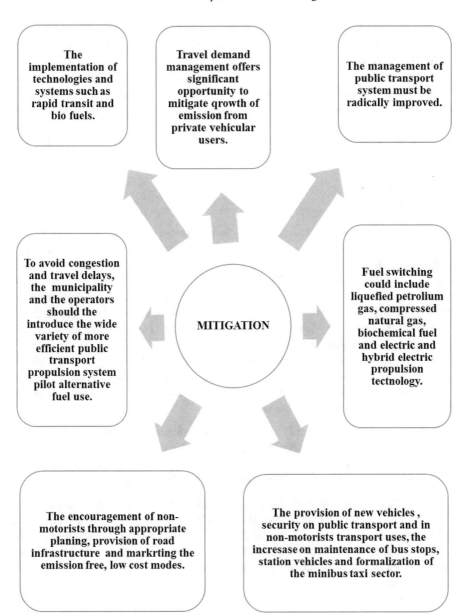

Fig. 4. Mitigation and strategies.

concerning the correlations between weather variables and traffic controls. Although, there is still a lack of theoretical understanding on how climate is perceived by road users and the affection of their travelling decisions and other traditional determinants such as travel cost and time (Keay and Simmonds 2006).

As the current trend suggested that by the year 2030 road traffic death will become the fifth leading cause unless action is taken or implemented (ASIRT 2014; TFTAHC 2000). It is reported that about 34% of the total accidents is due road environment and weather is considered one of the most important factors among those related to road environment. The re-evaluation road human factor such judgement, age, attention, driving skills, fatigue and experience should be considered. Also, the regular vehicular checkups need to be considered (Böcker et al. 2013). Road environmental aspects such as geometric alignments, grades, cross-sections, traffic control devices, surface friction, visibility and weather should be considered due to their effect on accidents.

3 Conclusion

The road users need to understand that the rainfall reduces visibility which even may be only a few meters during heavy rainfall, further reduced due to splashing water, particularly from heavy vehicles. The increased humidity during rainfall can cloud windows and windscreens which reduces visibility to see the other vehicles or an obstruction. The oncoming vehicles headlights reflection from the logged water on the roadway during heavy rainfall can also cause blinding during the night. Rainfall automatically reduces the friction of the road surface and can possibly lead to dynamic aquaplaning that can potentially increase road accidents.

This paper concludes that the rising population and increasing dependence on efficient reliable mobility have increased the importance of resilience to climate change. The standard authorities of climate change impact assessment (CIA) need an understanding of three important aspects such as:

- How climate change currently affects the road infrastructure and operations.
- How climate change may alter the frequency and magnitude of such impact.
- How concurrent technological and socio-economic development may shape the transportation system in future, exacerbating the effect of climate change.

The regular regression analysis or extrapolation of physical and behavioral relationship on the future climate change is a need to planning and implementation of a new development due to radical climate change. It is recommended that future climate change impact assessments should then focus on various key aspects such as the better representation of sub-daily extremes in climate tools and recreation of realistic spatially coherent climate.

References

Annual Global Road Crash Statistics: Association For Safe International Road Travel (ASIRT), Rockville, USA (2014)
Böcker, L et.al.: Impact of everyday weather on individual daily travel behaviours in perspective: a literature review. Transp. Rev. https://doi.org/10.1080/01441647.2012.747114
Brilon, W., Ponzlet, M: Variability of speed-flow relationships on German autobahns. Transp. Res. Rec. J. Transp. Res. Board (1996). https://doi.org/10.3141/1555-12

Brodsky, H., Hakkert, A.: Risk of a road accident in rainy weather. Accid. Anal. Prev. (1998). https://doi.org/10.1016/0001-4575(88)90001-2

Duranton, G., Turner, M.: the fundamental law of road congestion: evidence from US cities. Am. Econ. Rev. (2011). https://doi.org/10.1257/aer.101.6.2616

Field, B., et al.: Change I, Report C. 9781107654815: Climate Change 2014: Mitigation of Climate Change: Working Group III Contribution to the IPCC Fifth Assessment Report - AbeBooks - Intergovernmental Panel on Climate Change: 1107654815. Abebookscom 2018 (2014). https://www.abebooks.com/9781107654815/Climate-Change-2014-Mitigation-Working-1107654815/plp. Accessed 20 Mar 2018

Golob, T., Recker, W.: Relationships among urban freeway accidents, traffic flow, weather, and lighting conditions. J. Transp. Eng. (2003). https://doi.org/10.1061/(asce)0733-947x(2003) 129:4(342)

Goodwin, P., et al.: Elasticities of road traffic and fuel consumption with respect to price and income: a review. Transp. Rev. (2004). https://doi.org/10.1080/0144164042000181725

Griffiths, J., Rao, M.: Public health benefits of strategies to reduce greenhouse gas emissions. BMJ (2009). https://doi.org/10.1136/bmj.b4952

Hultkrantz, L., et al.: The value of improved road safety. Risk Uncertain. **32** (2006). https://doi. org/10.1007/s11166-006-8291-z

Jacobsen, P.: Safety in numbers: more walkers and bicyclists, safer walking and bicycling. Injury Prev. (2003). https://doi.org/10.1136/ip.9.3.205

Jaroszweski, D., et al.: The impact of climate change on urban transport resilience in a changing world. Prog. Phys. Geogr. (2014). https://doi.org/10.1177/0309133314538741

Liang, W., et al.: Effect of environmental factors on driver speed: a case study. Transp. Res. Rec. J. Transp. Res. Board (1998). https://doi.org/10.3141/1635-21

Liu, C.: Weather variability and travel behaviour – what we know and what we do not know. Transp. Rev. (2017). https://doi.org/10.1080/01441647.2017.1293188

Keay, K., Simmonds, I.: Road accidents and rainfall in a large Australian city. Accid. Anal. Prev. (2006). https://doi.org/10.1016/j.aap.2005.06.025

Kwon, T., et al.: Location optimization of road weather information system (RWIS) network considering the needs of winter road maintenance and the traveling public. Comput. Aided Civil Infrastruct. Eng. (2016). https://doi.org/10.1111/mice.12222

Mallick, D.: National climate change policy objective, strategy & action plans. SSRN Electron. J. (2015). https://doi.org/10.2139/ssrn.2584922

Marchini, J.: The Poisson Distribution (2008)

Maze, T., et al.: Whether weather matters to traffic demand, traffic safety, and traffic operations and flow. Transp. Res. Rec. J. Transp. Res. Board (2006). https://doi.org/10.3141/1948-19

Öörni, R., Kulmala, R.: Model for the safety impacts of road weather information services available to road users and related socio-economic benefits. IET Intell. Transp. Syst. (2012). https://doi.org/10.1049/iet-its.2011.0206

Shaw, C., et al.: Health co-benefits of climate change mitigation policies in the transport sector. Nat. Clim. Chang. (2014) https://doi.org/10.1038/nclimate2247

Takeuchi, M.: Snow accretion on traffic-control signs and its prevention. J. Japan. Soc. Snow Ice (1978). https://doi.org/10.5331/seppyo.40.117

Tiwari, G., Jain, D.: Accessibility and safety indicators for all road users: case study Delhi BRT. Transp. Geogr. (2012). https://doi.org/10.1016/j.jtrangeo.2011.11.020

Traffic Flow Theory and Highway Capacity. National Academy Press, Washington, D.C. (2000)

Wu, K., et al.: Developing a hierarchical structure of the co-benefits of the triple bottom line under uncertainty. J. Clean. Prod. (2018). https://doi.org/10.1016/j.jclepro.2018.05.264

Woodcock., J. et al. (2009). Public health benefits of strategies to reduce greenhouse-gas emissions: urban land transport. Lancet **374**(9705). https://doi.org/10.1016/s0140-6736(09)61714-1

Reaction Behaviour of Drivers to Road Markings: Case Study of Main South Road Lesotho – N8 Road South Africa

Jacob Adedayo Adedeji[1(✉)], Samuel Olugbenga Abejide[1],
Moliehi Monts'i[1], Mohamed Mostafa Hassan[1],
and Wafaa H. H. Mostafa[2]

[1] Sustainable Urban Roads and Transportation (SURT) Research Group,
Department of Civil Engineering, Central University of Technology,
Bloemfontein, Free State, Republic of South Africa
{jadedeji,mmostafa}@cut.ac.za, bskrtell@gmail.com,
moliehimontsi@gmail.com
[2] Faculty of Specific Education, Port Said University, Port Said, Egypt
nhwafaahussain@yahoo.com

Abstract. Road user behavior is one of the most critical components of traffic system, this is because it is unpredictable in nature. The variation in the behavioral responses of the road users has significantly contributed to the increase in traffic fatality rate. Consequently, the need for traffic control measures arises. Traffic control system such as traffic signals, traffic signs and road markings, on a road network tends to considerably reduce the number of conflicts and minimize road user's mistakes. However, the mortality rate on road accidents is on the increasing trend globally. Focusing on road markings as a traffic control system, the conducted study investigates the reaction behavior of drivers to road marking. Using random sampling methods, cross-border drivers between Main South-Road Lesotho and N8 Road, South Africa were interviewed on their experiences when driving on marked and unmarked roads. The questionnaire includes sections on driver personal experience, driver's reactions to road markings and the necessity for road marking and other traffic controls. Results show that approximately 67.7% of the drivers agree that their psychological state is influenced by road markings. Furthermore, using chi-square statistical analysis, results establish that the gender, age and educational background characteristics of drivers and their psychological responses to road marking are dependent on each other. In conclusion, emphasis on the necessity of road marking in reducing traffic fatality rate and the psychological effect of the unavailability of road marking on driver's behavior in most developing countries is presented.

1 Introduction

Road traffic fatality is at an alarming state, as there is a continuous increase in the accident rate annually. Despite the vast improvement in road transportation sectors, the number of fatalities remains high. Thus, according to WHO's report in 2014, traffic

© Springer Nature Switzerland AG 2019
S. El-Badawy and J. Valentin (Eds.): GeoMEast 2018, SUCI, pp. 58–71, 2019.
https://doi.org/10.1007/978-3-030-01911-2_6

fatalities in South Africa reached 11, 890 or 2.12% of total deaths (Road Traffic Management Corporation 2018) and Lesotho traffic fatalities in 2014 reached 486 or 1.79% of total deaths (Lehohla 2009). Overall, the global increase in road traffic mortality was predicted to be 67% by 2020 if appropriate action is not taken (Lehohla 2009). The major causes of traffic fatality have been attributed to the four critical traffic components which are; road user's characteristics, vehicle characteristics, road and its components (traffic furniture), and the environmental conditions. Nevertheless, of all the components, the road user's characteristics is the most dynamic in nature and unpredictable.

Furthermore, studies on traffic safety concluded that road user's characteristics are the main cause of accidents yet, it can be argued that it is not the only cause rather other components contribute to it (Niezgoda et al. 2012; Agbonkhese et al. 2013; Adedeji et al. 2016). However, all these components make road traffic a complex system, which can influence the probability of causing accidents (Niezgoda et al. 2012). Overall, if one of these critical traffic components is not well considered in road traffic, the impact will be felt by the road user's characteristics component, which will consequently lead to traffic fatality. Therefore, this study pertains to the reaction behavior of drivers as road user to road markings considering Main South-Road Lesotho and N8 Road South Africa as a case study.

1.1   Critical Traffic Component: Road User's Characteristics

Road user's characteristics are one of the most important traffic components, as traffic safety depends on the performance of the human factor to archive safety. According to Slinn et al. (2009) and Road Traffic Management Corporation (2017), 70% of road accidents are caused by human factors. The sole reason for this is the diversity in the behavior of road users and the diversity in the road user response to events. Driver's behavior can be measured through different means such as self-reported/expert-reported behaviors, vehicle measures of driving performance and even psychophysi-ological measures (Niezgoda et al. 2012). Furthermore, the driver expectancy phe-nomenon contributes to the reaction of the drivers and in a way affects their performances. Drivers expect things to operate in certain ways and when the expec-tancy is incorrect, either the driver takes longer to respond properly, or he/she may respond poorly or wrongly (Russell 1998). If, for example, a driver relies on the road marking for overtaking on any section of the road and now suddenly find him/herself on a road section without markings, one can imagine the difficulty the driver may have to experience in making decisions. Overall, the measure of the performance or behavior of drivers can be valid if all the basic components (e.g. traffic signals) of traffic are available on site.

1.2   Road Markings: Safety Features

Road marking, however, is any kind of device or material used on the road surface to convey formal information. These markings make a vital contribution to safety by clearly defining the path to be followed and are used as communication tools on paved roadways to provide guidance and information to the road users (Ozelim and Turochy

2014; Rehman and Duggal 2015). Road markings can be either mechanical (cat's eye, botts' dots, rumble strips and reflective markers), non-mechanical (paintings) or temporary in nature. As in other communication tools, road markings should be uniform to minimize confusion and uncertainty about their meaning and efforts exist to standardize such markings across borders (Mathew and Krishna Rao 2007; Rehman and Duggal 2015).

Furthermore, as aforementioned, road markings can fall in all the categories of communication tools such as: regulatory signs, warning signs, and informative signs. In the aspect of regulatory signs, the stop paint marking at the end of an intersection communicate to the road users (motorists) to stop for a while and observe other vehicles before progressing. Additionally, road-marking act as a warning sign in the case of rumble strips (also known as sleeper lines) which serve as noise generators, also attempts to wake a sleeping driver or alter a driver to various upcoming hazards by both sound and the physical vibration of the vehicle. Transverse road markings as defined by Martindale and Urlich (2010) are used to assist in raising driver awareness of risk through perceptual optical effects, thus encouraging drivers to reduce their speed in anticipation of an upcoming hazard.

Overall, road marking alerts the road users with information. In addition, road marking is found as one of the contributing factors to accident when poorly maintained (Elwakil et al. 2014) and a study based on the USA over a period of twenty years has shown that road markings can reduce fatality by 13% (FHWA 1994 cited in Grosskopf 2001). In South Africa, road markings are used on all paved road surfaces (Grosskopf 2001); However, this is not the case in Lesotho as the availability of road marking on roads is at fair to poor state. Nonetheless, the fairly available road markings on Lesotho roads are poorly maintained (Monts'i 2018) and thus, the markings fade off leaving the unmarked and prone to accidents.

2 Methodology and Data Collection

2.1 Study Area Description

In order to achieve the aim of this study, a case study area was selected purposefully as it consists of both the marked road surface and the unmarked road surface section, along a popular travel route. The Main South-Road Lesotho to N8 road South Africa, approximately 150 km (Fig. 1), connecting Lesotho to South Africa through Bloemfontein in Free State Province which also serves as one of the borders between Lesotho and South Africa.

Lesotho is enclosed fully by the Republic of South Africa in the Southern part of Africa. The country Lesotho falls amongst the poorest in the Southern part of Africa and largely depends on South Africa for its economy (Cobbe 1983). Most of the infrastructure take over mainly from funding developed countries or even world organizations such as The World Bank. Lesotho road network comprises of approximately 5,859 km, only 1,515 km is paved, while the others are unpaved and earth roads. However, the road network in South Africa is approximately 746,978 km, of which only 153,791 km is paved (The South Africa National Roads Agency Ltd.

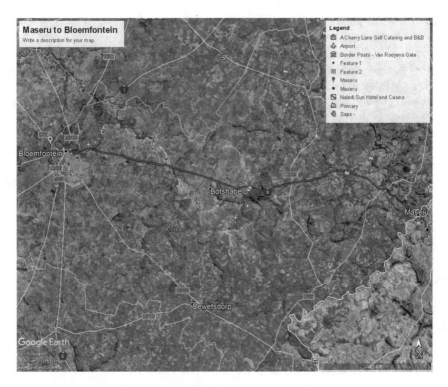

Fig. 1. Map showing route from Main South road Lesotho to N8 road South Africa (Google Earth 2018)

2010). About 90% of the paved roads in South Africa is flexible pavement (De Beer *et al.* 1999). The N1 and N8 are the basic national roads in Bloemfontein road network. The maintenance culture on the paved South Africa road can be rated a pass as the routine maintenance of the road marking are keep to date (Mostafa 2018), however, this is not the same in Lesotho (Monts'i 2018).

2.2   Data Collection

The approach used to collect data involved the use of questionnaires defined by sample size specifying a given margin of error, which will justify the population size of the total road users and provide a confidence interval of judgment (Gray 2013). The method used in this research involved the use of questionnaires randomly administered and completed by drivers traveling along the Main South road Lesotho and N8 road, South Africa. Two hundred questionnaires were distributed amongst drivers in various inter-city motor parks in Lesotho and Bloemfontein, however, only 127 questionnaires were returned.

The questionnaire used consists of three sections; the first section classifies the personal information (demographic characteristics) of the drivers: age, gender, educational level which influence their level of behavior and comprehension of the traffic signs and road markings, issuing country of driver's license and type. The second section of the questionnaires concerns the driver's experience and behavior while driving on road, this composed of multiple-choice questions with questions like; driving license disqualification, driving fines received, the driver's perception about his/her tendency to be distracted while he/she is driving, including the use of alcohol, energy drinks or medicines. The section also contains also questions related to the compliance to driving rules (seat-belt use, safety distance, speed limits, and rules of overtaking) and the driving style. The third section looks at road conditions and the availability of road markings and traffic signs using a Likert-type scale. Where the driver may indicate the consequences of the class using a 5-point scale (where 1 = never and 5 = always). Overall, seeks the opinions of the drivers on the further ways to improve road safety along the route.

The chi-square statistical tool was employed to test for the hypothesis relating to driver's characteristics. Thus, the null and alternative hypotheses for the testing were:

H_o: The driver's characteristics and psychological influence of road marking are independent of each other.

H_i: The null hypothesis is not true.

The expected cell frequencies were compared with the observed cell frequencies using the test chi-square, as estimated.

$$X^2 = \sum \frac{(O_i - E_j)^2}{E_{ij}}$$

where:

X^2 chi-square

O_{ij} observed frequency of the cell in the i th row and j th column

E_{ij} expected frequency of the cell in the i th row and j th column

The calculated chi-square result was compared with the critical chi-square value (using the table) with (r-1) x (c-1) degree of freedom to make a decision regarding the acceptance or rejection of the null hypothesis, Kothari (2004).

Decision Rule: If $X^2_{tab} > X^2_{cal}$, accept H_o otherwise reject

3   Evaluation and Interpretation of Results

3.1   Theme I: Demographic Information and Drivers Experience

The first section, whose results are summarized in Table 1, represents the approach between interviewers and interviewed drivers in the face-to-face survey. This section is characterized by having interviewed 14.9% of youngsters (age between 18 and 25 years), 59.8% between the age of 26 to 40 years, 23.6% between the age of 41 to 65 years, while 2.4% are over 65 years old. 35.4% of the interviewees are females and 64.6% are males and the education level ranges from 7.1% for Grade 1–6, 23.6% for

Grade 7–12, 44.1% for undergraduate and 25.2% are postgraduate, thus, it is expected of them to have a proper knowledge of road marking. Furthermore, majority of the drivers interviewed were from Lesotho (56.7%) while others are South African, and majority (67.7%) of the drivers are licensed for light motor vehicles (license type B), 27.6% for heavy motor vehicle (license type C) and the rest are motorcycles (license type A).

Table 1. Demographic characteristics of respondents

Demographic characteristics	Class	Percentage
Age	From 18 to 25	14.2%
	From 26 to 40	59.8%
	From 41 to 65	23.6%
	Above 65	2.4%
Gender	Female	35.4%
	Male	64.6%
Education	Grade 1–6	7.1%
	Grade 7–12	23.6%
	Undergraduate	44.1%
	Postgraduate	25.2%
Driver license issuer	South Africa	43.3%
	Lesotho	56.7%
License type	A	4.7%
	B	67.7%
	C	27.6%

The second section is about driver's behaviors and gives details of driver's personal experience and assessment (Table 2). Majority of the drivers interviewed have been driving for a while, with 11% of the drivers driving for over 23 years, 47.2% driving for 8 to 22 years and novice drivers about 41.7%. This results from the activities which the Lesotho people come to South Africa for, such as work, and academic studies. Nevertheless, 84.3% respondents have received fines for violation of traffic rules within the past one year and 40.9% respondents have been involved in car accident over the last 3 years respectively. This maybe as a result of the fact that majority at 73% of the drivers are within the age bracket of 18 to 40 years, thus, carefree driving style is associated to such age. Consequently, 86.6% of the respondents never faced driving license disqualification, this is as a result of the fact that this system was tested in few provinces and yet to be implemented in South Africa. However, this question might have been confused with the issue of failing driver's license test.

Furthermore, only 15% of the interviewed drivers are 'never' distracted while driving, although, over 60% of the drivers sometimes use their mobile phone while driving and take energy drink before driving. Consequently, the never distracted could be justified based on the 41.7% novice drivers. Most of the drivers interviewed drive over long distances and 62.2% psychological state are 'sometimes' stressed while

Table 2. Driver's experience and behavior

Driver's experience and behavior	Class	Percentage
Driving license disqualification	No	86.6%
	Yes	13.4%
Years of driving license	0 to 7	41.7%
	8 to 22	47.2%
	23 to 47	10.2%
	Above 47	0.8%
Fines received within the past 1 year	0 to 5	84.3%
	6 to 10	15.7%
Involved in a car accident within the last 3 years	No	59.1%
	Yes	40.9%
Distracted while driving	Always	3.9%
	Often	14.2%
	Sometimes	66.9%
	Never	15.0%
Using mobile phone while driving	Always	2.4%
	Often	7.1%
	Sometimes	69.3%
	Never	21.3%
Take energy drink before driving	Always	5.5%
	Often	17.3%
	Sometimes	44.9%
	Never	32.3%
Drive when stressed	Always	3.9%
	Often	15.0%
	Sometimes	62.2%
	Never	18.9%
Drive over long distances	Always	31.5%
	Often	29.1%
	Sometimes	33.1%
	Never	6.3%
Using Safety Belts	Always	51.2%
	Often	31.5%
	Sometimes	16.5%
	Never	0.8%
Respecting safety distance	Always	52.0%
	Often	23.6%
	Sometimes	24.4%
	Never	0%
Respecting speed limits	Always	44.9%
	Often	33.1%

(*continued*)

Table 2. (*continued*)

Driver's experience and behavior	Class	Percentage
		18.9%
	Never	3.1%
Respecting overtaking rules	Always	66.1%
	Often	20.5%
	Sometimes	13.4%
	Never	0%
Attention to passenger's seat belts	Always	29.9%
	Often	30.7%
	Sometimes	20.5%
	Never	18.9%

18.9% are often stressed when driving, these will have an impact on decision making for the drivers. Although, most of the drivers claimed 'always' to using safety belts (51.2%), respecting safety distance (52%), respecting overtaking rules *(dotted and the undotted white line markings on the travel route)* (66.1%), and minority pay attention to respecting speed limits (44.9%) and passengers seat belts (29.9%), these correlates to the *84.3%* receiving fines for violation of traffic rules.

3.2   Theme II: Road Conditions (Lesotho-Bloemfontein Route)

Section 3 investigate on the present road conditions along the Lesotho- Bloemfontein route. The first 3 set of question checked if the proposed target group were interviewed, results showed that 96.1% of the respondents use the route on weekly basis and only 3.9% respondents unsure of using the route in question. Furthermore, to evaluate the availability of road markings *(since it should reflect at night)* on this route the drivers were interviewed on their response and perception towards night driving, 92.1% respondents driving at night constantly and only 45.7% respondents often drive at night.

Overall, 59.8% respondents believe that the present road condition is good to excellent, 31.5% respondents states that the road in a fair condition and the other says it is in a poor to very poor state. Thus, this is validated by the availability of traffic signals and availability of road markings (Table 3). The drivers interviewed has a great understanding of the road marking, this is justified based on the standard process in obtaining the national driver's license in South Africa and Lesotho and also the educational level of the drivers interviewed. From the data collected, 66.9% respondents indicated that the available road markings are very useful, 70.9% respondents indicated that the drivers are not very comfortable while driving without the road markings and 87.4% respondents emphasized that road marking is a necessity and much more 81.1% responded that it is a necessity on road curves. Considering the usefulness, necessity, and comfortability of drivers driving on a marked road, it is eminent that road markings help in the process of decision making such as overtaking, maintaining once lane and obey the right of ways rules.

Table 3. Current road conditions along the case study area

Road conditions	Class	Percentage
Availability of traffic signals	Never Available	2.4%
	Fairly Available	3.9%
	Neutral	18.1%
	Available	49.6%
	Most Available	26.0%
Availability of road markings	Never Available	0.8%
	Fairly Available	3.1%
	Neutral	13.4%
	Available	40.9%
	Most Available	41.7%
Understanding of the road markings	Never Conversant	1.6%
	Fairly Conversant	0.8%
	Neutral	15.0%
	Conversant	32.3%
	Very Conversant	50.4%
Usefulness of road marking in making decisions	Never Useful	6.3%
	Fairly Useful	3.1%
	Neutral	7.1%
	Useful	16.5%
	Very Useful	66.9%
Comfortability level driving without road markings	Never Comfortable	70.9%
	Fairly Comfortable	11.8%
	Neutral	3.1%
	Comfortable	3.1%
	Very Comfortable	11.0%
Necessity of road makings	Never Necessary	0.8%
	Fairly Necessary	0%
	Neutral	2.4%
	Necessary	9.4%
	Very Necessary	87.4%
Necessity of road makings on curves	Never Necessary	0%
	Fairly Necessary	0%
	Neutral	1.6%
	Necessary	17.3%
	Very Necessary	81.1%

3.3 Theme III: Psychological Influence of Road Marking (PIRM) on Drivers

Figure 2 shows the psychological influence of road making on drivers, question such as; Do your fear level increase when road marking disappears? (PIRM1), Do you feel that your focus (concentration) on road increase when road marking disappears? (PIRM2) and Do you notice that other drivers are confused when there is no road marking? (PIRM3). Results for PIRM1 shows that 76.4% respondents fear level will increase, and this will affect their decision while driving, as the expectancy of the driver has been changed, thus, the driver will take a longer time to respond or respond poorly or wrongly. This is due to the fact that the ability of the drivers changes according to the speed of decisions making and their perfect timing. This is attributed to the theory of "cognitive neuroscience". In this case, a driver may take a decision based on his/her own conscious or as an influence of the other drivers (Walton *et al.* 2004). This will attribute to the length of the decision making which can negatively affect the other road users. Subsequently, that may affect the road safety and accident rates.

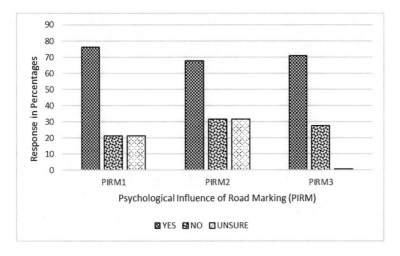

Fig. 2. Psychological influence of road marking on drivers

Additionally, 67.7% respondents agree that their concentration level will increase when the road marking disappears (PIRM2) and 31.5% respondents disagree, this in a way may cause unnecessary agitation in the decision making of the drivers. As it was explained earlier, the decision-making time can affect the other road users spreading the "*Amax-phobia*". This can happen at two levels; the first is fear of accidents while the second is fear of being afraid. This may attribute to more accidents rate due to loss of vehicle control. PIRM3, 70.9% respondents noticed that other drivers are confused when there is no road marking, 27.6% respondents are not confused and only 0.8% are unsure of the thoughts of the other drivers. This implies that the 70.9% are overthinking during the process of driving which can result in a poor decision or more errors. This high percentage further shows that road drivers are subject to "nervousness" as the road marking disappearance. This may cause heart beeps to increase and concentration to decrease which, in return, cause high mistakes rate as a parallel ratio.

3.4   Cross Analysis of Driver's Characteristics and Psychological Influence of Road Marking on Drivers

Evaluation of the three questions under the theme III was conducted to check the relationship between the psychological influence of road marking and individual characteristics with regards to gender, age, and educational background of the respondent drivers.

3.4.1   Cross-classification Analysis of Gender and Psychological Influence of Road Marking on Drivers

Table 4 shows the statistical result of the association between gender and the psychological influence of road marking. Considering 5% level of significance, the chi-square value is 5.99. However, the calculated chi-square values for all psychological influence of road marking on drivers were larger than the critical value, thus, indicating that there are some reasons to believe that the variables are dependent. Furthermore, since the individual X_{cal}^2 is greater than X_{tab}^2, reject H_o, meaning the gender characteristics of drivers and their psychological responses to road marking are dependent on each other.

Table 4. Cross analysis with gender of drivers

Gender	Psychological influence of road marking (PIRM) on drivers								
	PIRM1			PIRM2			PIRM3		
	YES	NO	UNSURE	YES	NO	UNSURE	YES	NO	UNSURE
MALE	61	18	1	55	26	1	57	23	2
FEMALE	36	9	0	31	14	0	33	12	0
Chi-square Test Statistic	116.84			85.58			93.56		
H_o Rejected?	Yes	Yes	Yes	Yes	Yes	Yes	Yes	Yes	Yes

3.4.2   Cross-classification Analysis of Age and Psychological Influence of Road Marking on Drivers

In the cross-classification analysis of age of drivers and psychological responses to road marking, the result is presented in Table 5. Similarly, considering a 5% level of significance, the chi-square value is 12.59 and the calculated chi-square values for all psychological influence of road marking on drivers were larger than the critical value. Therefore, indicating that there are some reasons to believe that the variables are dependent and thus, reject H_o. This implies that age characteristics and driver's psychological responses to road marking are dependent on each other.

Table 5. Cross analysis with age of drivers

Age	Psychological influence of road marking (PIRM) on drivers								
	PIRM1			PIRM2			PIRM3		
	YES	NO	UNSURE	YES	NO	UNSURE	YES	NO	UNSURE
18–25	15	3	0	11	7	0	14	4	0
26–40	58	17	1	54	22	0	54	21	1
41–65	22	6	2	20	9	1	20	9	1
Above 65	2	1	0	1	2	0	2	1	0
Chi-square Test Statistic	88.76			113.68			94.18		
H$_o$ Rejected?	Yes	Yes	Yes	Yes	Yes	Yes	Yes	Yes	Yes

3.4.3 Cross-classification Analysis of Educational Background and Psychological Influence of Road Marking on Drivers

Table 6 presents the cross-classification analysis of the educational background of the drivers and their psychological responses to road marking. Considering a 5% level of significance, the chi-square value is 12.59 and the calculated chi-square values for all psychological influence of road marking on drivers were larger than the critical value. Similarly, there are indications that the variables are dependent. Based on the value of the calculated chi-square greater than the critical value, the H$_o$ is rejected and implies that educational background and driver's psychological responses to road marking are dependent on each other.

Table 6. Cross analysis with educational background of drivers

Educational background	Psychological influence of road marking (PIRM) on drivers								
	PIRM1			PIRM2			PIRM3		
	YES	NO	UNSURE	YES	NO	UNSURE	YES	NO	UNSURE
Grade 1–6	4	3	2	7	1	1	4	3	2
Grade 7–12	26	3	1	21	9	0	18	12	0
Undergraduate	39	17	0	35	21	0	42	14	0
Postgraduate	28	4	0	23	9	0	26	6	0
Chi-square Test Statistic	121.47			88.61			94.19		
H$_o$ Rejected?	Yes	Yes	Yes	Yes	Yes	Yes	Yes	Yes	Yes

Overall, individual characteristic with regards to gender, age, and educational background have a great influence on the psychological responses of drivers to road markings.

4 Conclusions

Road markings and traffic signs are indispensable for road structure and disregarding them pose potential dangers to all road users. This study examines the behavioral response of driver to road marking. Results show that road markings are available along the case study area and the drivers understand the road marking, however, the application of the understanding is a key to the essence of these tools and this can lead to a reduction in traffic fatality. In this study, the necessity of road markings was also pointed out and the impact of the non-availability on driver's behavior was also noted. The study showed that in the absence of road markings the fear level of 76.4% of the drivers will increase and 67.7% will increase the concentration level, and this can cause unnecessary agitation in the decision making of the drivers. Therefore, the availability or non-availability of road marking has a great influence on the psychological state of mine of the drivers driving through a route and the individual characteristic with regards to gender, age, and educational background have a great influence on the psychological responses of drivers to road markings.

Overall, while developed countries continue with the improvement of vehicle characteristics, the stakeholders involved in the road designs and traffic regulations in South Africa, Lesotho, and other developing countries should focus on road conditions improvement by providing adequate resources for the maintenance of road to acceptable standards, more concertation of the inspection and maintenance (such as repainting of road marking) and enforcement of regulation and law should be stricter. Considering all these, the vision of traffic safety can be achieved in developing countries.

References

Adedeji, J.A., Abejide, S.O., Mostafa Hassan, M.: Effectiveness of Communication Tools in Road Transportation: Nigerian Perspective. In: Proceedings of the International Conference on Traffic and Transport Engineering ICTTE, Belgrade 2016, pp. 510–517 (2016)

Agbonkhese, O., Yisa, G.L., Agbonkhese, E.G., Akanbi, D.O., Aka, E.O., Mondigha, E.B.: Road traffic accidents in Nigeria: causes and preventive measures. Civil Environ. Res. 3(13), 90–99 (2013)

Cobbe, J.: The changing nature of dependence: economic problems in Lesotho. J. Mod. Afr. Stud. 21(2), 293–310 (1983)

De Beer, M., Kannemeyer, L., Fisher, C.: Towards improved mechanistic design of thin asphalt layer surfacing based on actual tyre/pavement contact stress-in-motion (SIM) data in South Africa. In: Proceedings of 7th Conference on Asphalt Pavements Southern Africa (1999)

Elwakil, E., Eweda, A., Zayed, T.: Modelling the effect of various factors on the condition of pavement marking. Struct. Infrastruct. Eng. 10(1), 93–105 (2014)

Gray, D.E.: Doing Research in the Real World. Sage, London (2013)

Grosskopf, S.E.: For safety's sake, let's do road marking quality control. In: Proceedings of the 20th South African Transport Conference South Africa, 16–20 July 2001

Kothari, C.R.: Research Methodology- Methods and Techniques, 2nd edn. New Age International Publishers, New Delhi (2004)

Lehohla, P.: Road Traffic Accident Deaths in South Africa, 2001–2006: Evidence from Death Notification. Statistics South Africa, Pretoria (2009)

Martindale, A., Urlich, C.: Effectiveness of transverse road markings on reducing vehicle speeds October 2010. NZ Transp. Agency Res. Rep. **423**(4) (2010)

Mathew, T.V., Krishna Rao K.V.: Chapter 36: Traffic signs (2007). Internet: http://nptel.ac.in/courses/105101008/28

Monts'i, M.: Use of life cycle assessment reference method to improve life span in Maseru roads. M. Tech dissertation, Central University of Technology, Free State (2018)

Mostafa Hassan, M.: Road maintenance: types and approaches in African countries. In: Proceeding of 2018 4th International Conference on Energy Materials and Environment Engineering (ICEMEE 2018), Malaysia, 13–15 April 2018

Niezgoda, M., Kamiński, T., Kruszewski, M.: Measuring driver behaviour-indicators for traffic safety. J. KONES **19**, 503–511 (2012)

Ozelim, L., Turochy, R.E.: Modelling retro reflectivity performance of thermoplastic pavement markings in Alabama. J. Transp. Eng. **140**(6), 05014001 (2014)

Rehman, S.A.U., Duggal, A.K.: Suitability of different material used for road marking-a review. Int. Res. J. Eng. Technol. **2**(2), 622–625 (2015)

Road traffic Management Corporation, South Africa: State of Road Safety Report January - December 2017 (2018). https://www.arrivealive.co.za/documents/RTMC%20Road%20Fatality%20Report%20for%202017.pdf. Accessed 25 June 2018

Russell, E.R.: Using concepts of driver expectancy, positive guidance and consistency for improved operation and safety. In: 1998 Transportation Conference Proceedings, pp. 155–158 (1998)

Slinn, M., Matthews, P., Guest, P.: Traffic Engineering Design, Principles and Practice, 2nd edn. Elsevier Butterworth-Heinemann, Oxford (2005)

The South Africa National Roads Agency Ltd.: Annal report 2010. [Internet] [Cited 5 December 2014] (2010). http://www.nra.co.za/content/Sanral_ARep_10_Web.pdf

Walton, M.E., Devlin, J.T., Rushworth, M.F.: Interactions between decision making and performance monitoring within prefrontal cortex. Nat. Neurosci. **7**(11), 1259 (2004)

Resilient Characteristics of Asphalt Stabilized Soil

Saad Issa Sarsam[1(✉)] and Aya Tawfeaq Kais[2]

[1] Department of Civil Engineering, Transportation Engineering,
University of Baghdad, Baghdad, Iraq
saadisasarsam@coeng.uobaghdad.edu.iq
[2] Department of Civil Engineering, College of Engineering,
University of Baghdad, Baghdad, Iraq

Abstract. The resilient behavior of asphalt stabilized subgrade soil in terms of changes in the deformation and shear failure under repeated loading was investigated in this work. Asphalt stabilized soil specimens of two sizes (100 mm diameter and 70 mm height) and (152 mm diameter with 127 mm height) have been prepared in the laboratory and compacted to its maximum dry density at optimum fluid requirement (water + liquid asphalt) and at 0.5% of fluid above and below the optimum. Specimens have been subjected to curing, then tested for deformation and resilient modulus under repeated shear stresses. The deformation was captured along the load repetition process with the aid of linear variable differential transformer (LVDT) under controlled stress and environmental conditions in the pneumatic repeated load system (PRLS) until failure. The large size specimens were tested under single punching shear stress, while small size specimens were tested under double punching shear stress after eight days of curing. The resilient deformation data of the two testing techniques under single and double punching shear stress was analyzed and compared. It was observed that the Double punching shear exhibit higher resilient strain than that of single punch shear in a range of (22, 8, and 24)% for (16.5, 16, and 15.5)% of fluid content respectively. The resilient modulus decreases after 1200 and 1800 load repetitions by (48, 47, 61, and 32)% and (50, 48, 63, and 34)% for (untreated soil, 15.5, 16, and 16.5)% fluid content respectively as compared to that after one load repetition under single punch shear stress. It was concluded that asphalt stabilization exhibit positive impact on resilient modulus. It increases by a range of three and seven folds under single and double punch shear stress after one load repetition by the addition of asphalt. Higher asphalt content exhibit reduction of Mr. Significant reduction in total and resilient strain could be detected after liquid asphalt was implemented in the subgrade soil.

Keywords: Asphalt · Double punching · Shear stress
Stabilization · Resilient modulus

© Springer Nature Switzerland AG 2019
S. El-Badawy and J. Valentin (Eds.): GeoMEast 2018, SUCI, pp. 72–84, 2019.
https://doi.org/10.1007/978-3-030-01911-2_7

1 Introduction

The stability of subgrade soil against horizontal displacement is essential for pavement design and construction, (Sarsam et al. 2016). Usually, proper compaction and proper gradation of subgrade soil exhibit stable platform for pavement and commonly do not exhibit need for treatment. However, if there is a deficiency in the gradation of the soil, mechanical stabilization could solve such problem by mixing it with another binding material, (Sarsam and Mohsen 2017). The subgrade soil layer is usually capable for resisting tensile and shear stresses and have sufficient stiffness to resist vertical and lateral deformation. Fill materials for subgrade construction may not meet these necessities, (Sarsam and Tawfeaq 2017) so the soil stabilization of subgrade soil with liquid asphalt could be an alternative to increase soil adhesion, stability, and improve the overall properties which are preferred for embankment construction objectives, (Sarsam et al. 2015). The mechanism of treatment with asphalt material include reduction of water penetration and adding cohesive strength to the soil by the asphalt; waterproofing action, aeration technique, cementation action (Jones et al. 2010). Soil stabilization with liquid asphalt can develop its geotechnical properties such as durability, strength, permeability, and compressibility, (Makusa 2012). Asphalt stabilization is considered as a sustainable step towards roadway construction on problematic subgrade soil, (Sarsam et al. 2017a). (Al-Daffaee 2002) stated that shear strength tests presented that the cohesion had improved with increasing asphalt content although the angle of internal friction reduced. It was concluded that using asphalt as a material for stabilization must be limited to reasonable concern with Gypseous soils. A detailed investigation on asphalt stabilization has been conducted by (Sarsam et al. 2017b). It was found that cohesion, C, and angle of internal frication (ϕ) of asphalt stabilized soil have increased with increasing asphalt content up to optimum, then the soil mechanical properties start to decease. The asphalt binder is responsible of the cohesive soil property and binds the soil elements by the thin film of liquid asphalt. It also possess some flexible properties and exhibit elastic action against deformation under repeated loading, (Sarsam and Barakhas 2015). (Taha et al. 2008) used cut-back asphalt (RC-70) for Gypseous soil stabilization and examined the shear strength of soil. The outcome of this study shows that at 6% asphalt cut-back content the shear strength of undrained condition (Cu) increased by 33% and the shear parameter increased until reach an optimum content of asphalt and then start to decrease with added more than optimum asphalt content. Cohesion increase with the increase in binder content by 70% while the angle of internal friction was not affected, beyond that it decreased. An experimental study was conducted by (Sarsam 1986) on asphalt stabilization using the California bearing ratio test CBR. It was reported that the CBR increases by 40% after asphalt stabilization, while the deformation decreases by 69% when the soil asphalt stabilization has been performed. As reported by (AL-Khayat 2010), the cut-back asphalt added to Gypseous soil generates a type of elastic features and strain rebound in the stabilized soil mixture at great stress application in the dry test condition. Deformation of asphalt stabilized embankment model under cyclic loading has been investigated by (Sarsam et al. 2014), it was stated that the shear properties of untreated soil are inadequate, whereas the addition of asphalt emulsion possess a mild increase in the shear properties for soaked and dry testing conditions. Implementation of 17% of emulsion has improved the shear strength properties by (12 and 14) folds, while the cohesion increased by (9 and 30) in the soaked and dry test conditions respectively as compared to untreated soil. (Siddiki and Kim 2006) stated that, in general, the resilient modulus of subgrades is influenced by the Deviator stress, the technique of compaction and dry density, and water content.

The aim of this investigation is to evaluate the resilient properties of asphalt stabilized soil under repeated shear stresses by assessing the deformation of soil-asphalt mixture under load repetitions. Repeated Single and double punching shear stresses will be implemented.

2   Materials and Method

2.1   Liquid Asphalt

Medium curing Cut-back asphalt of grade MC-30 obtained from Dora refinery was implemented, Table 1 shows its important physical properties which was obtained from the refinery.

Table 1. Medium curing cut-back asphalt MC-30 properties as per (ASTM, 2009)

Property	Test result
Grade	MC-30
Viscosity (C.st.) @ 60 °C	30–60
Flash point (Cleveland open cup) °C (minimum)	38
Water % by volume (max)	0.2
Distillation test to 360 °C, %Volume of total distillate: To 225 °C (maximum) To 260 °C (maximum) To 315 °C (maximum)	25 40–60 75–93
Test on residue from distillation	
Penetration @ 25 °C (100gm, 5 s., 0.1 mm)	120–250
Ductility @ 25 °C (cm)	+100
Solubility in Tricolor ethylene % weight (minimum)	990

2.2   Subgrade Soil

The subgrade soil was obtained from Al-Taji quarry, which is 25 km to the north of Baghdad as illustrated in Fig. 1. The grain size distribution of the tested soil is shown in Fig. 2. The physical properties of the subgrade soil are summarized in Table 2.

2.3   Preparation of the Asphalt Stabilized Specimens

Two sizes of specimens have been prepared. Specimens of (100 mm diameter and 70 mm height) and (152 mm diameter with 127 mm height) have been prepared in the laboratory and compacted to a target density of 1.76 gm/cm^3 and optimum fluid

Fig. 1. Location of subgrade soil

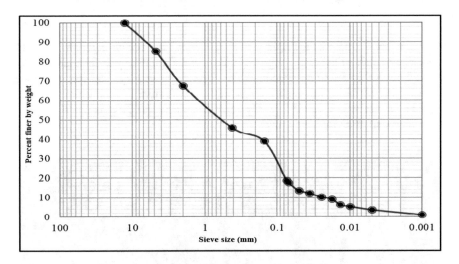

Fig. 2. Grain size distribution of subgrade soil

content of 16%, (10% water +6% liquid asphalt) and at 0.5% of fluid above and below the optimum. The soil was oven dried at 100 °C to a constant weight then cooled at room temperature of 25 °C. Soil was sieved through sieve No. 4, and the portion passing the sieve was implemented in this investigation. Specimens were prepared for each fluid content percentage as well as for the pure soil. Water was added to the soil and mixed by hand till the soil mixture became homogenous. At that moment, the required liquid asphalt percentage was added and mixed for 5 min at 25 °C until the asphalt soil mixture become consistent and all the particles were coated with asphalt. The soil asphalt mixture was spread in a tray and left 2 h for aeration at room temperature. Aeration might give greater density and strength, lower water absorption

Table 2. Physical properties of subgrade soil

Physical properties	Test result
Specific gravity	2.62
Liquid limit	23
Plastic limit	Non-Plastic
Plasticity index	Non-Plastic
Maximum dry density (Modified Compaction) (gm/cm^3)	1.760
Optimum moisture content (%)	16
AASHTO classification system	A-1-b
(ASTM, 2009) classification system	SM
Coefficient of curvature (C_C)	60
Coefficient of uniformly (C_U)	0.26
Cohesion C (kPa)	18
Angle of internal frication ($\varnothing$)	41
Un-drained shear strength (Unconfined compression test) S_U (kPa)	233
Shear strength using unconsolidated un-drained (UU) Triaxial test (kPa)	54

meanwhile the decreases in weight would be lower than before aeration, (Al-Nadaf 2013). Figure 3 show the aeration process. The aerated mixture was then placed inside the mold and subjected to compaction to meet its target density. The specimens were left for curing at 25 °C for eight days at room temperature before testing as recommended by (Sarsam and Kais 2017). Figure 4 illustrates part of the prepared specimens during compaction and curing.

Fig. 3. Aeration of asphalt mixture and curing of specimens

2.4 Specimens Testing

The pneumatic repeated load system (PRLS) shown in Fig. 5 has been implemented for the test. Repeated Uniaxial compressive loading was applied. The load excitation function is a rectangular wave with steady frequency equal to (1) cycles of 10 Hz per second. A maximum load pulse time is (0.1) s and (0.9) s is the rest period used throughout the interval of test. Specimens have been stored in the test chamber at laboratory temperature of 25 °C for two hours. Before starting test, the pressure

Fig. 4. Compaction and curing of CBR specimens

actuator was regulated to the specific pressure level of (55.16 kPa). LVDT was used to measure the deformation every second (every cycle) till the test is completed. The large size specimens (in the CBR mold) were subjected to repeated single punching shear stress from top surface, then from the bottom surface and the average value was considered for analysis as recommended by (Sarsam 1986). On the other hand, the small size specimens (in the Marshal mold) were subjected to repeated double punch shear stress.

3   Results and Discussion

Resilient modulus Mr was described as the ratio of applied deviator stress to recoverable or resilient strain. Resilient modulus is a measure of materials responses to load and deformation. Generally, higher modulus indicates greater resistance to deformation. Mr reflects the nonlinearity of asphalt stabilized soil. Using a regression analysis for power mathematical model. The resilient strain is a function calculated based on the resilient deformation obtained under constant stress load repetitions. Thus, the equation is illustrated as follows:

$$\varepsilon_r = \frac{rd}{h} \tag{1}$$

Where

ε_r = the resilient strain
rd = the resilient displacement
h = the specimen's height.

$$Mr = \frac{\sigma}{\varepsilon_r} \tag{2}$$

Where

Mr = Resilient modulus (MPa)
σ = Applied stress
ε_r = Resilient strain.

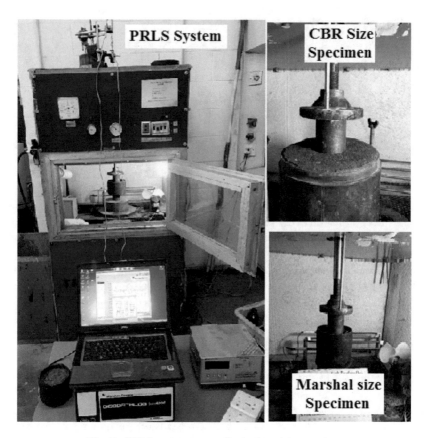

Fig. 5. PRLS and test setup for both specimens sizes

According to (Yoder and Witczak 1975), the power regression model is recommended to be used for the calculation of resilient modulus. Therefore, the model adopted for this study is based on the following equation

$$\varepsilon_r = aN^b \tag{3}$$

Where

ε_r = the resilient strain
(a) = Equation constant representing the intercept (resilient strain after one repetition)
(b)= Equation constants representing the slope (rate of change in resilient strain as a function from (N)
N= Number of load repetitions to failure.

3.1 Resilient Strain Under Single Punching Shear Test

California bearing ratio specimens have been implemented and tested using PRLS test system. Constant stress was applied throughout the time period of the test under 1800 repetition. Figure 6 shows the relation between the three types of strain (Total, permanent, and resilient) versus number of loading repetitions for repeated single punching shear on specimens after 8 days curing. The results show a non-linear behavior of deformation. It is shown that any further increase in the fluid content above 15.5%, will have a negative impact on deformation. When the fluid content increases, it will increase the flexibility. Thus, the resilient displacement enhanced, and permanent deformation will be decreased. However, the soil stabilized with cutback asphalt exhibit material properties of elastoplastic since the resilient strain represent the elastic term and permanent strain illustrate the plastic term. Figure 6 demonstrate a general decrease in total and resilient deformation comparative to a natural soil. Therefore, when asphalt film is coating the soil particle it will give some elastic properties, while there is no substantial influence on permanent deformation. Thus, the addition of asphalt to the soil within the optimum requirements will not have a significant impact on plastic properties of soil. Figure 7 demonstrate that the intercept decreases by (68, 62 and 29)% for (15.5, 16, and 16.5)% fluid content respectively as compared to the untreated soil condition. On the other hand, the slope increases by (4, 41.5, and 41)% for (15.5, 16, and 16.5)% fluid content respectively as compared to the untreated soil condition. This behavior may be attributed to the increase in cohesion between soil particles due to adhesion between soil particles and asphalt in addition to the particle interlock due to the gradation of the soil. The intercept (a) represents the permanent strain after the first load cycle (N = 1), where N is the number of the load cycles. The higher the value of intercept, the larger the strain and hence the larger the potential for permanent deformation as mentioned in the literature, (Sarsam and Mohsen 2017). However, slope (b) represents the rate of change in the permanent strain as a function of the change in loading cycles (N) in the log-log scale. High slope values for a mix indicate an increase in the material deformation rate hence less resistance against rutting. A mix with a low slope value is preferable as it prevents the occurrence of the rutting distress mechanism at a slower rate, (Sarsam and Barakhas 2015).

3.2 Resilient Strain Under Double Punching Shear Test

Marshal size specimens after eight days of curing period have been tested using PRLS test machine. The constant stress was applied throughout the time period of the test in terms of double punching shear stress. The maximum number of repetitions applied was 1800. The repeated load has been applied with constant excitation frequency of 10 Hz per second. Two confined specimens for each fluid content (total specimens are eight) has been tested Fig. 8 illustrated the displacement versus no. of repetitions for untreated and asphalt stabilized soil under double punch shear. Significant reduction in total and resilient strain could be detected after liquis asphalt was implemented in the subgrade soil. On the other hand, Fig. 9 exhibit reduced intercept values and increased slope after asphalt stabilization. The intercept decreases by (88, 87 and 71)% for (15.5, 16, and 16.5)% fluid content respectively as compared to the untreated soil condition,

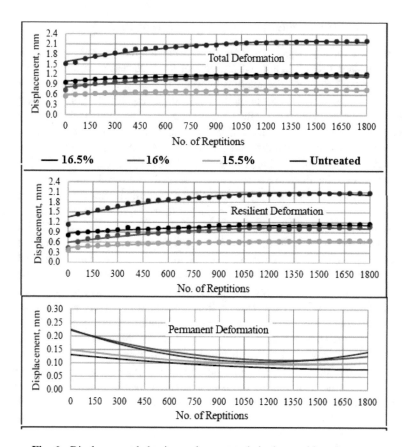

Fig. 6. Displacement behavior under repeated single punching shear test

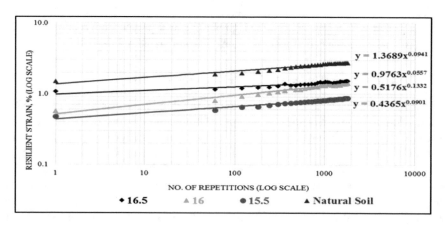

Fig. 7. Resilient strain verses fluid content under repeated single punch shear

while the slope increases by (277, 336, and 58)% for (15.5, 16, and 16.5)% fluid content respectively as compared to the untreated soil condition.

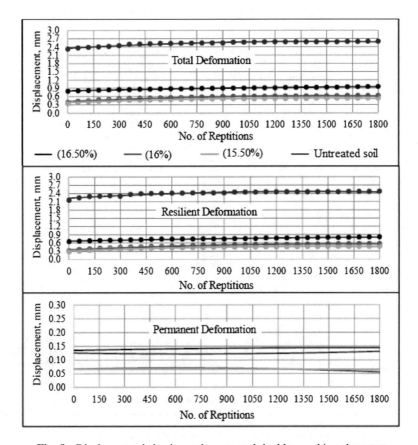

Fig. 8. Displacement behavior under repeated double punching shear test

Figure 10 present a comparative assessment of resilient strain among single and double punching shear application, it shows a comparison between the CBR and Marshal Size specimens at 1200 load repetitions. This number represent the borderline between macro and microdamage according to (Sarsam and Mohsen 2017). In addition, the results show an improvements in strain resistance for several types of deformations. It can be observed that double punching shear exhibit higher resilient strain than that of single punch shear in a range of (22, 8, and 24)% for (16.5, 16, and 15.5)% of fluid content respectively. Such behaviour may be attributed to the fact that loading from top and bottom on the specimens will increase the confinement state and increase the resilient strain when the load was removed.

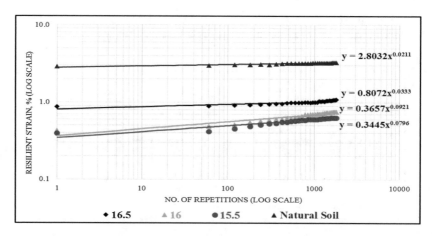

Fig. 9. Resilient strain verses fluid content under repeated double punch shear

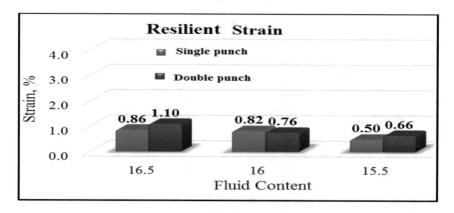

Fig. 10. Comparison of resilient strain among single and double punch shear

3.3   Resilient Modulus of Asphalt Stabilized Soil

Table 3 presents the resilient modulus Mr as calculated from the mathematical models after various load repetitions. It can be observed that Mr decreases as the load repetitions increases which indicate the start of failure for the untreated soil. Implementation of cutback asphalt had a positive impact on Mr, it increases by a range of three and seven folds under single and double punch shear stress after one load repetition by the addition of asphalt. Higher asphalt content exhibit reduction of Mr. The single punch shear exhibit lower Mr after one load repetition as compared to double punch while a significant higher Mr difference is detected at higher load repetition. The resilient modulus decreases after 1200 load repetitions by (48, 47, 61, and 32)% for (untreated soil, 15.5, 16, and 16.5)% fluid content as compared to that after one load repetition. Resilient modulus further decreases after 1800 load repetitions by (50, 48, 63, and 34)% for (untreated soil, 15.5, 16, and 16.5)% fluid content as compared to that after one load repetition. The double punch shear test exhibit similar behavior. It was felt that the loading of specimen from top and bottom at the same time as the case of double punch and the restricted thickness of the specimen may restrict the rolling process of soil particles over each other which is further restricted by the asphalt film surrounding each soil particle. This could restrict the deformation of the specimen in the vertical and horizontal directions and create high resilient modulus as compared to the case of single punch shear test.

Table 3. Resilient modulus after various load repetitions

Testing technique	Loading repetitions	Mr (MPa) for several fluid content, %			
		Untreated soil	15.5%	16%	16.5%
Single punch shear	1	40.3	126.5	106.6	56.5
Double punch shear	1	19.7	139.1	137.9	117.3
Single punch shear	1200	20.7	66.8	41.5	38.1
Double punch shear	1200	17.0	91.1	78.6	54
Single punch shear	1800	19.9	64.4	39.3	37.2
Double punch shear	1800	16.8	89.7	75.8	51.2

4   Conclusions

Based on the limited testing program, the following conclusions may be drawn:

1. The intercept decreases by (68, 62 and 29)%, while the slope increases by (4, 41.5, and 41)% for (15.5, 16, and 16.5)% fluid content respectively as compared to untreated soil condition under single punch shear stress after 8 days curing.
2. The intercept decreases by (88, 87 and 71)%, while the slope increases by (277, 336, and 58)% for (15.5, 16, and 16.5)% fluid content respectively as compared to untreated soil condition under double punch shear stress after 8 days curing.
3. Double punching shear exhibit higher resilient strain than that of single punch shear in a range of (22, 8, and 24)% for (16.5, 16, and 15.5)% of fluid content respectively.
4. The resilient modulus decreases after 1200 load repetitions by (48, 47, 61, and 32)% for (untreated soil, 15.5, 16, and 16.5)% fluid content as compared to that after one load repetition under single punch shear stress.
5. Resilient modulus further decreases after 1800 load repetitions by (50, 48, 63, and 34)% for (untreated soil, 15.5, 16, and 16.5)% fluid content as compared to that after one load repetition under single punch shear stress.

References

Al-Daffaee, A.H.: The Effect of Cement and Asphalt Emulsion Mixture on the Engineering Properties of Gypseous Soils. M.Sc. thesis, Civil Engineering Department, University of Al-Mustansiria, Iraq (2002)

AL-Khayat, B.: Implementation of Gypseous Soil-Asphalt Stabilization Technique for Course Construction. M.Sc. thesis, Civil Engineering Department, University of Baghdad (2010)

Al-Nadaf, H.Q.: Effect of Water Absorption on the Behavior of Asphalt Stabilized Gypseous Soil. M.Sc. thesis, Civil Engineering Department, University of Baghdad (2013)

American Society for Testing and Materials, ASTM: Road and Paving Material, Vehicle-Pavement System, Annual Book of ASTM Standards, vol. 04.03 (2009)

Jones, D., Rahim, A., Saadeh, S., Harvey, J.: Guidelines for the Stabilization of Subgrade Soils in California. Pavement Research Centre, University of California, California (2010)

Makusa, G.P.: Soil Stabilization Method and Materials. Luleå University of Technology, Luleå (2012)

Sarsam, S., Al Saidi, A., Al Taie, A.: Assessment of shear and compressibility properties of asphalt stabilized collapsible soil. Appl. Res. J. 2(12), 481–487 (2016)

Sarsam, S.I., Kais, A.T.: Comparative assessment of shear strength under static loading for asphalt stabilized soil. Int. J. Sci. Res. (IJSR), 6(11), November 2017

Sarsam, S., Mohsen, A.: Influence of additives and durability cycles on the resilient modulus of asphalt stabilized soil. J. Geotech. Eng. 4(2), 32–39p (2017)

Sarsam, S.I., Tawfeaq, K.A.: Assessing the variation in shear properties of asphalt stabilized soil using three testing techniques. Int. J. Eng. Pap. IJEP 2(2), 1–12 (2017)

Sarsam, S.I., Husain, A., Almas, S.: Assessing the variation in cutback asphalt requirement for various geotechnical properties of asphalt stabilized soil. J. Geotech. Eng. 2(3), 17–29 (2015). STM Journals

Sarsam, S., Al Saidi, A., Al Taie, A.: Influence of combined stabilization on the structural properties of subgrade soil. J.Geotech. Eng. 4(1), 13–24 (2017a). STM Journals

Sarsam, S., Al Saidi, A., Jasim, A.: Monitoring of the compressibility characteristics of asphalt stabilized subgrade. Int. J. Sci. Res. Knowl. 5(1), 011–019 (2017b)

Sarsam, S., Barakhas, S.: Assessing the structural properties of asphalt stabilized subgrade soil. Int. J. Sci. Res. Knowl. IJSRK 3(9), 227–240 (2015)

Sarsam, S.: A study on California bearing ratio test for asphaltic soils. Indian Highways IRC, vol. 14, No. 9, India (1986)

Sarsam, S.I., Alsaidi, A.A., Alzobaie, O.M.: Impact of asphalt stabilization on deformation behavior of reinforced soil embankment model under cyclic loading. J. Eng. Geol. Hydrogeol. JEGH 2(4), 46–53 (2014). Sciknow publication Ltd. USA

Siddiki, N., Kim, D.: Simplification of Resilient Modulus Testing for Subgrades. Technology Transfer and Project Implementation Information (2006)

Taha, M.Y., Al-Obaydi, A.H., Taha, O.M.: The use of liquid asphalt to improve gypseous soils. Al-Rafidain Eng. 16(4), 38–48 (2008)

Yoder, E.J., Witczak, M.W.: Principles of Pavement Design. Wiley, New York (1975)

Analytical Study of Headway Time Distribution on Congested Arterial: A Case Study Palestine Road in Baghdad City

Zainab Ahmed Alkaissi[(⊠)]

Highway and Transportation Department,
AlMustansiriyah University, Baghdad, Iraq
zainablkaisi77@googlegmail.com

Abstract. The time headway is important parameter in traffic flow theory and broadly applied in different branches in transportation engineering. In this research the one of the major arterial street in Baghdad city is Palestine arterial street that have been selected as a case study to investigate the distribution of time headway under heavy flow conditions. Collected field data for tow links; Link (1) from Al-Mawal intersection to Bab AlMutham intersection 1.03 km length and link (2) from Zayona intersection to Mayslone intersection is obtained at two different time periods to conform the variation of time headway under congestion periods. The variation in time headway for Link (1) is reduced and more constant state is obtained due heavy flow conditions at congestion peak periods. The situation is not the same on Link (2), the fluctuation in time headway is still observed at higher flow rate. This demonstrated the effect of land use characteristics on time headway distribution and vehicle arrivals. Both Shape for the distribution of probability density function is skewed to the right and the peak rises for link (1) and its higher than for link (2). This illustrated that the peak value for probability density function is higher on link (1) which implies high flow rate at congestion periods. The Logostic probability distribution function is used to probably describe the time headway at congested Palestine arterial street for link (1) and (2). The fitted field data of time headway for different range of scale parameter from 1.5 to 0.85 are obtained. Goodness of fit using Chi Square non- parametric test is applied in this research to explore how the theoretical Logostic distribution fits the empirical data for time headway distribution.

Keywords: Time headway · Statistical analysis · Urban arterial
Congested arterial

1 Introduction

In traffic engineering applications the time headway properties of vehicles are very important task in the studies of highway capacity, level of service, unsignalized intersection and roundabouts. Also the traffic simulation is depend on the modeling of vehicle time headway since its represents an important characteristics in microscopic flow theory that influence on the driver behavior and road safety.

© Springer Nature Switzerland AG 2019
S. El-Badawy and J. Valentin (Eds.): GeoMEast 2018, SUCI, pp. 85–97, 2019.
https://doi.org/10.1007/978-3-030-01911-2_8

Many studies based on the modeling of headway time distribution, Al-Ghambi (2001), Al-Ghambi (1999) investigated time headway under low, medium and high flow rates and he concluded that at arterial street the time headway follows Gamma distribution and Erlang distribution especially on high flow rates. Abtahi et al. (2011) notified that there was different time headway distribution for passing and middle lanes in urban highways for heavy traffic conditions and suggest that the lognormal and Gamma distribution are best fit with shifting parameter of 0.24 s and 0.69 s for passing and middle lanes respectively. Edigbe et al. (2014) reported that the overtaking induces minimal headways and the Pearson type III and Erlang distribution are fit the headway distribution frequency.

Sara (2014) stated that an appropriate models are selected for headway distribution for heavy vehicles and passenger cars at different flow rates using Chi-Square test. It was conformed that the variation of time headway distribution for heavy vehicles and passenger cars is attributed to the different behavior of drivers under congestion in the vicinity of heavy vehicles and passenger cars.

Another earlier researchers have proposed the distribution models for time headway, such as, Buckley (1968) semi-Poisson model, Cowan (1975) M3 distribution model, the log-normal distribution (Mei and Bullen 1999) and the double displaced negative exponential distribution (Sullivan and Troutbeck 1994). Zhang and Wang (2013) reported that Cowan's M3 model is widely applied due to its simplicity and easy approximations for describing longer headways.

2   Study Area and Data Collected

One of the major arterial street in Baghdad city is Palestine arterial street that have been selected for this research. Its majority evolved from the mix land use characteristics; commercial, residential and educational surrounding area and the location in the east of Baghdad running parallel to the west of Army Cannal between Al-Mustansiriyah Square through Beirut square to the end of it at Maysalone square. Palestine arterial street consists of 6 divided lane carriageway 3-lane in each direction. Figures 1and 2 present the study area of Palestine arterial street with the region of selected links.

Two links parts of major Palestine arterial street in Baghdad city were chosen in this research Link (1) from Al-Mawal intersection to Bab AlMutham intersection 1.03 km length and link (2) from Zayona intersection to Mayslone intersection as mention in previous section. Three days take up for field data collection on 23 Monday, 24 Tuesday and 25 Wednesday 2016 which have been considered in this research. To conform the variation of time headway two time periods during the evening peak hour 6:00–6:30 pm to 6:30–7:00 pm are investigated. Relies on these time period, the time headway data are classified depending upon the flow rate conditions. A four set of field data is collected of 100 data point each with totally 400 data point. The weather conditions during field data collection is of good visibility and sunny with adequate and appropriate distance from the adjacent traffic signalized intersections.

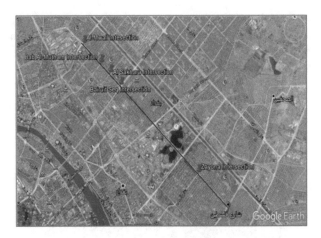

Fig. 1. Study area of Palestine arterial street.

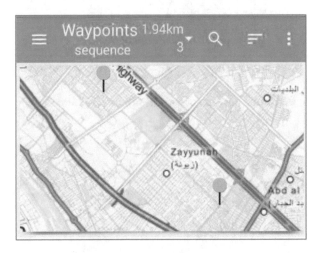

Fig. 2. Palestine arterial street with selected two links region.

3   Results and Discussions

3.1   Headway Time Distribution

Based on the collected data is for two links of Palestine arterial street; Link (1) from Al-Mawal intersection to Bab AlMutham intersection and link (2) from Zayona intersection to Mayslone Intersection; at two different time periods 6:00–6:30 pm with flow rate 1042 vph and 6:30–7:00 pm with flow rate 1308 vph for link (1) and for link (2) a flow rate of 1104 vph at time period 6–6:30 pm and 1164 vph at 6:30–7:00 pm as explained in Table 1. The two selected links were chosen in this research to compromise the effect of commercial and residential region on vehicle arrival. The frequency distributions for time headway with normal curve are estimated using SPSS (SPSS ver.21 statistical software) as shown in Figs. 3 and 4 respectively. The statistical variation in mean, medial, mode and variance also presented in Table 1. For links (1) and (2) at different flow rates for different time periods at peak hour.

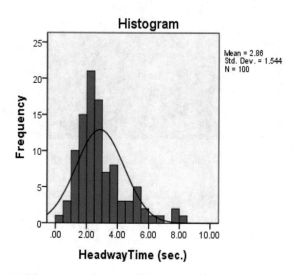

(a) Time Period 6:00-6:30 pm at 1042 vph flow rate.

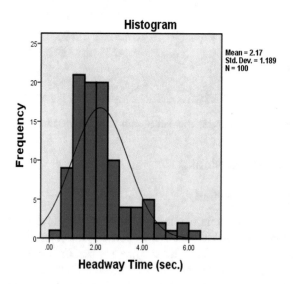

(b) Time Period 6:30-7:00 pm at 1308 vph flow rate.

Fig. 3. Time headway distribution for link (1).

Link (1) which represents commercial urban area at Palestine arterial street in Baghdad city attained a mean value of headway time about 2.86 s and standard deviation of 1.544 s at flow rate 1042 vph, and decreased to 2.17 s of mean time headway and 1.189 s of standard deviation at flow rate 1308 vph.

Table 1. Descriptive statistics and flow rate at different time periods for link (1) and link (2).

	Link (1)		Link (2)	
	6:30–7:00 pm	6:00–6:30 pm	6:30–7:00 pm	6:00–6:30 pm
Flow rate (vph)	1042	1308	1104	1164
Mean	2.86	2.17	2.52	2.5
Median	2.45	2.106	2.1	2
Mode	1.9	1.48	2.1	2
Standard deviations	1.544	1.189	1.633	1.655
Variance	2.384	2.191	2.668	2.738
Sample size	100	100	100	100

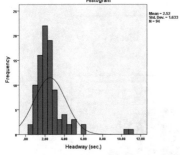
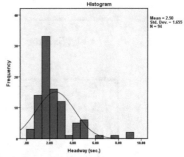

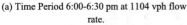

(a) Time Period 6:00-6:30 pm at 1104 vph flow rate.

(b) Time Period 6:30-7:00 pm at 1164 vph flow rate.

Fig. 4. Time headway distribution for link (2).

This is predictable since the variation in time headway is reduced and more constant state is obtained due heavy flow conditions at congestion peak periods which appears in Fig. 3b, the more uniform shape of frequency distribution. The situation is not the same on Link (2), the fluctuation in time headway is still observed at higher flow rate and this can referred to the residential region as seen in Fig. 4a and b. This demonstrated the effect of land use characteristics on time headway distribution and vehicle arrivals.

3.2   Probability Density Distribution

The Probability density function is estimated for both link (1) and (2) as shown in Figs. 5 and 6 respectively. Both Shape for the distribution of probability density function is skewed to the right and the peak rises for link (1) and its higher than for link (2). This illustrated that the peak value for probability density function is higher on link (1) which implies high flow rate at congestion periods.

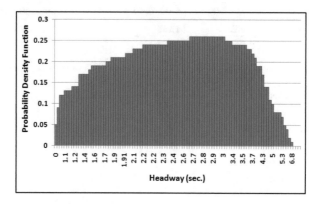

Fig. 5. Probability density function for link (1).

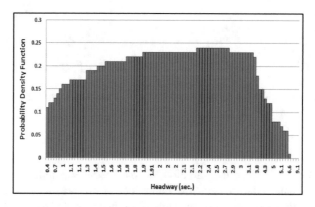

Fig. 6. Probability density function for link (2).

3.3   Time Headway Statistical Modeling

The Probability density function is estimated for both link (1) and (2) as shown in Figs. 7 and 8 respectively. Both Shape for the distribution of probability density function is skewed to the right and the peak rises for link (1) and its higher than for link (2). This illustrated that the peak value for probability density function is higher on link (1) which implies high flow rate at congestion periods.

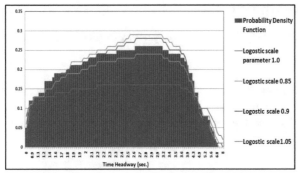

(a) Time headway Distribution 6:00- 6:30pm.

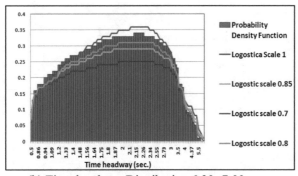

(b) Time headway Distribution 6:30- 7:00pm.

Fig. 7. Probability density function of time headway distribution for link (1) at different scale parameter.

Table 2 presents the best fit for time headway distribution with probability density function of Logostic distributions with scale parameter of 0.85 and 1.05 for link (1) at time periods 6:00–6:30 pm and 6:30–7:00 pm and scale parameter of 1.1 for link (2) at both time periods respectively; which illustrated the significant level of p -value greater than 0.05.

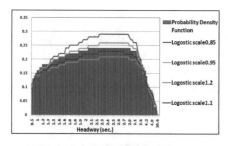

(a) Time headway Distribution 6:00- 6:30pm.

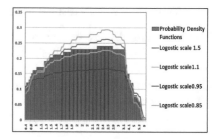

(b) Time headway Distribution 6:30- 7:00pm.

Fig. 8. Probability density function of time headway distribution for link (2) at different scale parameter.

For the validation of the Logostic probability distribution function of time headway additional filed data for time headway were collected. A comparison between the observed field data and theoretical time headway are shown in Figs. 9 and 10 for link (1) and link (2) respectively.

Table 2. Test statistics of the estimated scale parameter for logostic density function distribution for link (1) and link (2).

			Kolmogorov-Smirnov[a]			Shapiro-Wilk		
			Sig.	df	Statistic	Sig.	df	Statistic
Link (1)	Time headway distribution 6:00– 6:30 pm	Logostic distribution scale parameter 0.85	.000	99	.181	.000	99	.852
	Time headway Distribution 6:30– 7:00 pm	Logostic distribution scale parameter 1.05	.000	99	.175	.000	99	.848
Link (2)	Time headway distribution 6:00– 6:30 pm	Logostic distribution scale parameter 1.1	.000	99	.170	.000	99	.839
	Time headway distribution 6:30– 7:00 pm	Logostic distribution scale parameter 1.1	.000	99	.212	.000	99	.804

A good fit is obtained for link (1) and (2) as shown in Figs. 11 and 12 respectively.

Goodness of fit using Chi Square non- parametric test is applied in this research to explore how the theoretical Logostic distribution fits the empirical distribution for time headway. An additional field data is obtained and divided into different time periods 6:00–6:30 pm and 6:30–7:00 pm. As shown in Table 3 the Chi-Square goodness of fit test is less than the table value then the null hypothesis is accepted and enhanced that there is no significant difference between the observed and theoretical value for time headway.

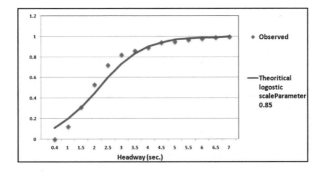

(a) Cumulative Probability of Time headway
Distribution 6:00- 6:30pm.

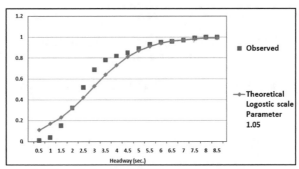

(b) Cumulative Probability of Time headway
Distribution 6:30- 7:00pm.

Fig. 9. Observed and theoretical time headway data for link (1).

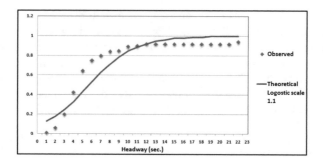

(a) Cumulative Probability of Time headway
Distribution 6:00- 6:30pm.

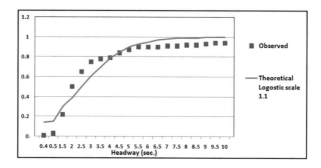

(b) Cumulative Probability of Time headway
Distribution 6:30- 7:00pm.

Fig. 10. Observed and theoretical time headway data for link (2).

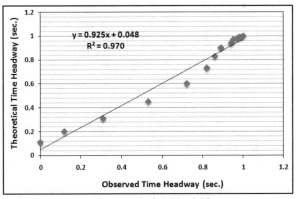

(a) At Time Period 6:00- 6:30pm.

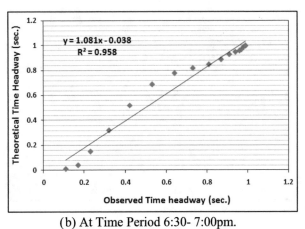

(b) At Time Period 6:30- 7:00pm.

Fig. 11. Theoretical versus observed time headway for link (1).

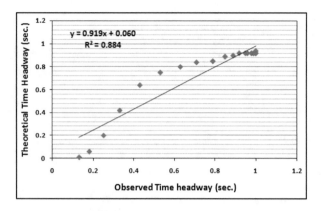

(a) At Time Period 6:00- 6:30pm.

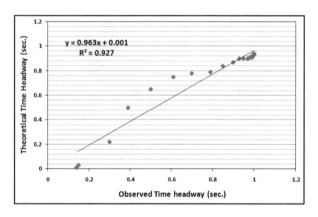

(b) At Time Period 6:30- 7:00pm.

Fig. 12. Theoretical versus observed time headway for link (2).

Table 3. Chi Square goodness of fit for link (1) and link (2) at different time periods at significant level $\alpha = 0.05$.

	Time period	N	df	χ^2	$\chi_{critical}$
Link (1)	6:00–6:30	17	16	0.335	26.296
	6:30–7:00	15	14	0.193	23.685
Link (2)	6:00–6:30	20	19	0.538	30.144
	6:30–7:00	20	19	0.399	31.41

4 Conclusions

This research displayed the analysis of time headway distribution for selected urban arterials in Baghdad city. The following remarks points are drawn:

1. A mean value of headway time about 2.86 s and standard deviation of 1.544 s at flow rate 1042 vph is obtained for link (1) and decreased to 2.17 s of mean time headway and 1.189 s of standard deviation at flow rate 1308 vph due the variation in time headway is reduced and more constant state is obtained due heavy flow conditions at congestion peak periods. On the other hands the situation is not the same on Link (2), the fluctuation in time headway is still observed at higher flow rate and this can referred to the residential region. This demonstrated the effect of land use characteristics on time headway distribution and vehicle arrivals.
2. The Shape of probability density distribution function is skewed to the right for link (1) and link (2) but the peak for link (1) is higher than for link (2) which implies high flow rate at congestion periods.
3. The Logostic probability distribution function is used to probably describe the time headway at congested Palestine arterial street at link (1) and (2) for scale parameter range from 1.5 to 0.85 for link (1) at time periods 6:00–6:30 pm and 6:30–7:00 pm and scale parameter of 1.1 for link (2) at both time periods respectively.
4. There is no significant difference between the observed and theoretical value for time headway using the Logostic distribution model. Chi Square non- parametric test goodness of fit is applied in this research to explore how the theoretical Logostic distribution fits the empirical data of time headway distribution.

References

Al-Ghamdi, S.: Analysis of time headways on urban roads: case study from Riyadh. J. Transp. Eng. ASCE **127**(4), 289–294 (2001)

Al-Ghamdi, A.S.: Modeling vehicle headways for low traffic lows on urban freeways and arterial roadways. In: Proceeding of the 5th International Conference on Urban Transport and the Environment for the 21st Century, Rhodes, Greece (1999)

Abtahi, S.M., Tamannaei, M., Haghshenash, H.: Analysis and modelling time headway distributions under heavy traffic flow conditions in the urban highways: case of Isfahan. Transport **26**(4), 375–382 (2011)

Buckley, D.J.: A semi-poisson model of traffic flow. Transport. Sci. **2**(2), 107–133 (1968)

Cowan, R.: Useful headway models. Transport. Res. **9**, 371–375 (1975)

Edigbe, J., et al.: Headway distributions based on empirical Erlang and Pearson type III time methods compared. Res. J. Appl. Sci. Eng. Technol. **7**(21), 4410–4414 (2014)

Mei, M., Bullen, G.R.: Lognormal distribution for high traffic flows. Transportation Research Record 1398, TRB, National Research Council, Washington, D.C., pp. 125–128 (1999)

Sullivan, D.P., Troutbeck, R.J.: The use of Cowan's M3 headway distribution for modeling urban traffic flow. Traffic Eng. Control **35**(7), 445–450 (1994)

Sara, M.: Evaluating the time headway distributions in congested highways. J. Traffic Logist. Eng. **2**(3) (2014)

Zhang, G., Wang, Y.: A Gaussian kernel-based approach for modelling vehicle headway distributions. Transp. Sci. **48**(2), 206–216 (2013)

Attenuation Effect of Material Damping on Impact Vibration Responses of Railway Concrete Sleepers

Sakdirat Kaewunruen[1]([⊠]), Ange-Theres Akono[2], and Alex M. Remennikov[3]

[1] Railway and Civil Engineering, School of Engineering, The University of Birmingham, Birmingham, UK
s.kaewunruen@bham.ac.uk
[2] Civil Engineering, Northwestern University, Chicago, IL, USA
[3] School of Civil, Mining and Environmental Engineering, University of Wollongong, Wollongong, NSW, Australia

Abstract. In railway industry, high strength concrete has been adopted for track slabs and railway sleepers for more than half a century. Prestressed concrete sleepers (or railroad ties) are designed usually using high strength concrete (>55 MPa) in order to carry and transfer the wheel loads from the rails to the ground and to maintain rail gauge for safe train travels. In general, the railway sleepers are installed as the crosstie beam support in ballasted railway tracks. Statistically, they are subjected to impact loading conditions induced by train operations over wheel or rail irregularities, such as flat wheels, dipped rails, crossing transfers, rail squats, corrugation, etc. These defects can be commonly found during the operational stage of life cycle. The magnitude of the shock load depends on various factors such as axle load, types of wheel/rail imperfections, speeds of vehicle, track stiffness, etc. This paper demonstrates the investigations into the dynamic responses of *in-situ* prestressed concrete sleepers using high strength materials, particularly under a variety of impact loads. The nonlinear finite element model of full-scale prestressed concrete sleeper with the realistic support condition has been developed using a finite element package, STRAND7. It has been verified by the experiments carried out using the high capacity drop-weight impact machine and experimental modal testing. The experimental results exhibited very good correlation with numerical simulations. In this paper, the numerical studies are extended to evaluate the dynamic behaviors of high strength concrete sleepers modified by crumb rubbers to increase material damping coefficients. The outcome of this study can potentially lead to the utilization and practical design guideline of high strength concrete engineered by crumb rubber from wasted tires and plastics for prestressed concrete sleepers.

© Springer Nature Switzerland AG 2019
S. El-Badawy and J. Valentin (Eds.): GeoMEast 2018, SUCI, pp. 98–107, 2019.
https://doi.org/10.1007/978-3-030-01911-2_9

1 Introduction

Without a doubt, the majority of civil infrastructure is constructed out of concrete, currently at a rate of 2 billion tonnes per year (Chung 1995). This is somehow responsible for 5% of global carbon dioxide emissions annually (Kaewunruen et al. 2017). However, it is well known that concrete has several disadvantages such as low tensile strength, low ductility, brittle, low damping (low energy dissipation), and high susceptibility to cracking. This causes the structure to deteriorate and lose its integrity when subjected to repeated harsh environmental conditions and dynamic loading conditions (Remennikov and Kaewunruen 2008; Kaewunruen 2014; Meesit and Kaewunruen 2017). Thus, when exposed to these high-intensity conditions, concrete structures are at a risk of failure. In addition, the high global usage of concrete material combined with the large amount of pollution its production produces every year is a major concern. Paris Agreement in 2016 has imposed the limit of carbon emission so that global warming can be limited to be less than 2 °C in 2100 (Binti Sa'adin et al. 2016a, 2016b). This implies that the use of high-carbon materials such as cement should be even more efficient and effective as much as possible. Therefore a sustainable approach needs to be taken to find solution to these existing issues in material production and selection for design and manufacturing (Kaewunruen and Lee 2017). The sustainable approach within this study involves developing a method to reduce carbon emissions and to improve the resilience of concrete structures. This study comprises of novel concrete innovation incorporating waste materials (see Fig. 1) for the purposes of reducing carbon emissions and also improving damping of concrete (Kaewunruen and Kimani 2017; Kaewunruen et al. 2018a).

Fig. 1. Waste car tyres for recycling

A vital safety-critical component of railway track structures is railway sleepers (also called 'railroad tie' in North America). Railway sleepers are the cross beam element supporting rails in order to provide load support and to secure rail gauge. Today, the most common material for manufacturing sleepers is concrete (Kaewunruen et al. 2014; You et al. 2017). The experience of design and application of railway concrete sleepers have been over 60 years around the world. Their key functions are to redistribute loads from the rails onto the underlying ballast bed, and to secure rail gauge for safe and smooth train passages. Based on the current design approach using static and quasi-static theory of solid mechanics, the design life span of the concrete sleepers is targeted at around 50 years in Australia and around 70 years in Europe (Standards Australia 2003; British Standards Institute 2016). In design practice, dynamic problems have not fully been taken into account, giving rise to the lack of new innovation for concrete sleepers. Current industry practice is still based on the topological optimisation using static analysis and the selection of tailored or bespoke dynamic factors for quasi-static design (Remennikov et al. 2012; Kaewunruen and Remennikov 2009; Wolf et al. 2015; Vu et al. 2016). This is because the current design and testing standards are rather primitive and overly simplified. Figure 2 shows a typical ballasted railway tracks. The track superstructure includes rail, rail pads, fasteners, sleepers and ballast; and the track substructure contains ballast mat, subballast (or capping layer), geosynthetics, subgrade and formation.

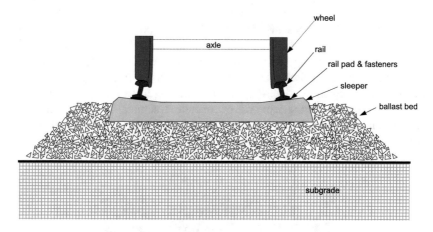

Fig. 2. Typical ballasted track

Despite the fact that the railway sleepers are exposed to dynamic loading conditions (Remennikov and Kaewunruen 2008), the material damping aspect has never been fully investigated. This paper is the first to present an advanced railway concrete sleeper modeling capable of analysis into the vibration attenuation effects of dynamic loading on the dynamic behaviors of railway concrete sleepers. The emphasis of this study is placed on the nonlinear dynamic flexural responses of railway concrete sleepers subjected to effective viscous damping of concrete material. It is the first time that the responses of concrete sleepers incorporating material damping have been investigated. The insight into the vibration attenuation will help structural and track engineers making a better choice in advanced material design and selection. It will also inspire materials engineers to further improve the dynamic material capabilities.

2 Nonlinear Modelling

A number of extensive studies recommended that the two-dimensional Timoshenko beam model is the most suitable option for modeling concrete sleepers under vertical loads (Neilsen 1991; Cai 1992; Grassie 1995; Manalo et al. 2012). In this investigation, the finite element model of *in-situ* concrete sleeper (optimal length of 2.5 m) has been previously developed and calibrated against the numerical and experimental modal parameters (Kaewunruen and Remennikov 2006a, 2006b, 2008a, 2008b). Figure 3 illustrates the two-dimensional finite element model for an in-situ railway concrete sleeper. Using a general-purpose finite element package STRAND7 (G+D 2001), the numerical model included the beam elements, which take into account shear and flexural deformations, for modeling the realistic concrete sleeper. The trapezoidal cross-section was assigned to the sleeper elements. The rails and rail pads at railseats were simulated using a series of spring. In this study, the sleeper behaviour is stressed so that very small stiffness values were assigned to these springs. In reality, the ballast support is made of loose, coarse, granular materials with high internal friction. It is often a mix of crushed stone, gravel, and crushed gravel through a specific particle size distribution. It should be noted that the realistic ballast provides resistance to compression only (Kaewunruen and Remennikov 2008b).

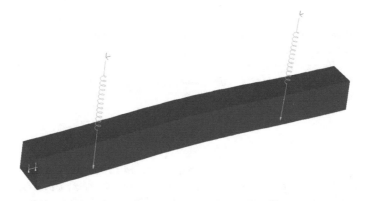

Fig. 3. STRAND7 finite element model of a prestressed concrete sleeper

In this study, the realistic support condition is simulated using the tensionless beam support feature in STRAND7. This attribute allows the beam to lift or hover over the support while the tensile supporting stiffness is omitted. This attribute creates nonlinear boundary conditions to the sleeper model, requiring Newton Raphson's numerical iterations to resolve the sleeper-ballast contact perimeter. The tensionless support

Table 1. Engineering properties of the standard sleeper used in the modeling

Parameter lists		
Flexural rigidity	$EI_c = 4.60$, $EI_r = 6.41$	MN/m^2
Shear rigidity	$\kappa GA_c = 502$, $\kappa GA_r = 628$	MN
Ballast stiffness	$k_b = 13$	MN/m^2
Rail pad stiffness	$k_p = 17$	MN/m
Sleeper density	$\rho_s = 2{,}750$	kg/m^3
Sleeper length	$L = 2.500$	m
Rail gauge	$g = 1.435$	m
Wheel load distance	$d = 1.500$	m

option can correctly represent the ballast characteristics in real tracks. Table 1 shows the geometrical and material properties of the finite element model. It is important to note that the parameters in Table 1 give a representation of a specific rail track. These data have been validated and the verification results have been presented elsewhere (Kaewunruen and Remennikov 2006a, 2008a).

In structural design, it is common to assume that concrete material has negligible viscous damping ratio (Hesameddin et al. 2015). However, it is often found that the effective damping of high-strength concrete can be varied from 0.1% to 2% (Meesit and Kaewunruen 2017). In general, the equation of forced motion for multi-degree-of-freedom (MDOF) system can be generalised. Taking a mass-normalized formulation approach, the acceleration measured on the structure becomes the mass-normalized inertial force. The mass-normalized velocity proportional equivalent viscous damping ratio can be written in terms of the damping coefficient and the natural period of a sub-critically structure as (Kaewunruen et al. 2018a):

$$\beta = \frac{c}{m} \frac{T_n}{4\pi} \tag{2}$$

where β is the equivalent viscous damping ratio; m is the mass of structure; c is equivalent viscous damping; and T_n is the natural period of the structure (i.e. period of the dominate mode of response). In this study, the first bending mode of sleeper vibration has been used to calculate the damping ratio. The equivalent viscous damping (c) has been used in the FE modelling for transient dynamic integration in order to avoid matrix errors for explicit finite solution calculations.

3   Results and Discussion

Free vibration analysis has been conducted to evaluate the natural frequencies and corresponding mode shapes of *in-situ* concrete sleeper. Figure 4 shows the dominant bending modes of vibration of the sleepers. As a result, Tn is 7 ms and the mass of sleeper (m) is 354 kg.

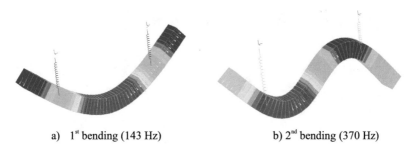

a)   1st bending (143 Hz)   b) 2nd bending (370 Hz)

Fig. 4. Free vibration characteristics of prestressed concrete sleeper

The dual wheelset impact loads of 100 kN magnitude and 3 ms duration are applied at both railseats to stimulate impact vibrations. This impulse is equivalent to the effect of common wheel burns (e.g. 3–5 mm flats) on railway tracks. The effects of material damping on the impact responses of railway concrete sleeper at railseats and at mid-span can be illustrated in Fig. 5. It is clear that material damping affects the impact responses across the frequency span. The higher the frequency the higher the loss of impact spectra. The damping of concrete can suppress well the impact vibrations at both railseats and mid-span of the concrete sleeper. This can be implied that the improvement in material damping can considerably suppress vibrations that can cause breakage of sleeper and underlying ballast. This insight can also be observed for railway bridge viaducts (Ülker-Kaustell and Karoumi 2012; Zhai et al. 2013; Malveiro et al. 2018; Kaewunruen et al. 2018b). The dynamic load effects can be suppressed, resulting in lesser dynamic defections and bending stresses. Since the concrete sleepers are generally designed to be '*uncracked*' under serviceability limit state (i.e. dynamic impact factor of 2.0 to 2.5), the results clearly show that damping enhancement (>4% of damping ratio) can significantly improve the long-term performance and durability of the concrete sleepers.

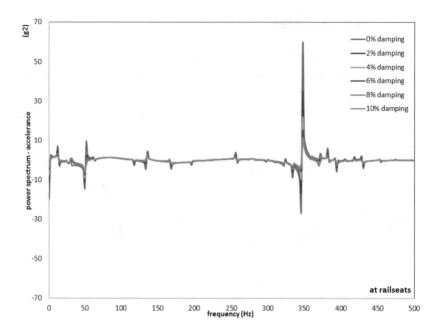

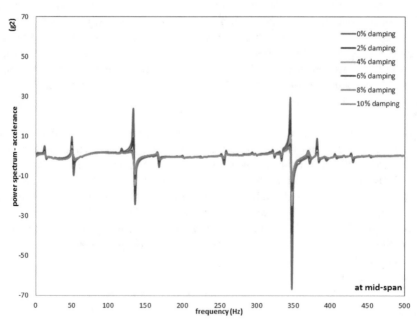

Fig. 5. Impact responses of railway concrete sleeper (frequency domain)

4 Conclusions

The most cost-effective railway track systems for metro, light rail, urban and suburban and freight networks are the ballasted tracks. For ballasted tracks, concrete is the most common used material for railway concrete sleepers. For over five decades, concrete sleepers are used to redistribute wheel forces onto track structure and secure track gauge for safe passages of rolling stocks. Statistically, most of the time, railway tracks experience dynamic load conditions. For design purpose, concrete material's damping characteristic is often neglected. Thus, the understanding into vibration attenuation of the sleeper due to the material damping is rather limited. The ignorance of damping has resulted in very little research into advanced concrete technology for railway applications. This study is the world first to incorporate advanced knowledge of novel concrete with high damping for dynamic design of railway concrete sleepers. This paper highlights the effects of concrete damping on the vibration attenuation of railway concrete sleepers in a track system. Using an established and validated finite element model of concrete sleepers, realistic sleeper-ballast contact conditions have been adopted for nonlinear transient analysis. This study is the first to reveal that the concrete damping can provide high level of vibration attenuation in concrete sleepers in a track system across wide range of frequencies. This insight will help structural and track engineers to make a better choice of advanced concrete and composite materials for manufacturing railway concrete sleepers.

Acknowledgments. The first author gratefully acknowledges the Japan Society for the Promotion of Science (JSPS) for his JSPS Invitation Research Fellowship (Long-term), Grant No L15701, at Track Dynamics Laboratory, Railway Technical Research Institute and at Concrete Laboratory, the University of Tokyo, Tokyo, Japan. The authors would also like to thank British Department for Transport (DfT) for Transport – Technology Research Innovations Grant Scheme, Project No. RCS15/0233; and the BRIDGE Grant (provided by University of Birmingham and the University of Illinois at Urbana Champaign). Finally, the authors are sincerely grateful to the European Commission for the financial sponsorship of the H2020-RISE Project No. 691135 "RISEN: Rail Infrastructure Systems Engineering Network" (www.risen2rail.eu), which enables a global research network that tackles the grand challenge of railway infrastructure resilience and advanced sensing in extreme environments (Kaewunruen et al. 2016).

References

Binti Sa'adin, S.L., Kaewunruen, S., Jaroszweski, D.: Risks of climate change with respect to the Singapore-Malaysia high speed rail system. Climate **4**, 65 (2016a). https://doi.org/10.3390/cli4040065

Binti Sa'adin, S.L., Kaewunruen, S., Jaroszweski, D.: Heavy rainfall and flood vulnerability of Singapore-Malaysia high speed rail system. J. Aust. J. Civil Eng. **14**(2), 123–131 (2016b). https://doi.org/10.1080/14488353.2017.1336895

British Standards Institute: European Standard BS EN13230 Railway applications. Track. Concrete sleepers and bearers, London, UK (2016)

Cai, Z.: Modelling of rail track dynamics and wheel/rail interaction, Ph.D. thesis, Department of Civil Engineering, Queen's University, Ontario, Canada (1992)

Chung, D.: Strain sensors based on the electrical resistance change accompanying the reversible pull-out of conducting short fibers in a less conducting matrix. Smart Mater. Struct. **4**(1), 59–61 (1995)

G+D Computing. Using Strand7: Introduction to the Strand7 finite element analysis system, Sydney, Australia (2001)

Hesameddin, P.H., Irfanoglu, A., Hacker, T.J.: Effective viscous damping ratio in seismic response of reinforced concrete structures. In: 6th International Conference on Advances in Experimental Structural Engineering, 1–2 August 2015. University of Illinois, Urbana-Champaign, United States (2015)

Grassie, S.L. Dynamic modelling of concrete railway sleepers. Journal of Sound Vibr. **187**, 799–813 (1995)

Kaewunruen, S.: Impact damage mechanism and mitigation by ballast bonding at railway bridge ends. Int. J. Railw. Technol. **3**(4), 1–22 (2014). https://doi.org/10.4203/ijrt.3.4.1

Kaewunruen, S., Lee, C.K.: Sustainability challenges in managing end-of-life rolling stocks. Front. Built Environ. **3**, 10 (2017). https://doi.org/10.3389/fbuil.2017.00010

Kaewunruen, S., Kimani, S.K.: Damped frequencies of precast modular steel-concrete composite railway track slabs. Steel Compos. Struct. Int. J. **25**(4), 427–442 (2017)

Kaewunruen, S., Remennikov, A.M.: Sensitivity analysis of free vibration characteristics of an in-situ railway concrete sleeper to variations of rail pad parameters. J. Sound Vibr. **298**(1), 453–461 (2006a)

Kaewunruen, S., Remennikov, A.M.: Nonlinear finite element modeling of railway prestressed concrete sleeper. In: Proceedings of the 10th East Asia-Pacific Conference on Structural Engineering and Construction, EASEC 2010, vol. **4,** pp. 323–328 (2006b)

Kaewunruen, S., Remennikov, A.M.: Effect of a large asymmetrical wheel burden on flexural response and failure of railway concrete sleepers in track systems. Eng. Fail. Anal. **15**(8), 1065–1075 (2008a)

Kaewunruen, S., Remennikov, A.M.: Experimental simulation of the railway ballast by resilient materials and its verification by modal testing. Exp. Tech. **32**(4), 29–35 (2008b)

Kaewunruen, S., Remennikov, A.M.: Influence of ballast conditions on flexural responses of railway concrete sleepers. Concr. Aust. J. Concr. Inst. Aust. **35**(4), 57–62 (2009)

Kaewunruen, S., Remennikov, A.M., Murray, M.H.: Introducing a new limit states design concept to railway concrete sleepers: an Australian experience. Front. Mater. **1**, 8 (2014). https://doi.org/10.3389/fmats.2014.00008

Kaewunruen, S., Sussman, J.M., Matsumoto, A.: Grand challenges in transportation and transit systems. Front. Built Environ. **2**, 4 (2016). https://doi.org/10.3389/fbuil.2016.00004

Kaewunruen, S., Rachid, A., Goto, K.: Damping effects on vibrations of railway prestressed concrete sleepers. In: World Multidisciplinary Civil Engineering-Architecture-Urban Planning Symposium, IOP Conference Series: Material Science and Engineering (2018a, accepted)

Kaewunruen, S., Ishida, T., Remennikov, A.M.: Dynamic performance of concrete turnout bearers and sleepers in railway switches and crossings. Adv. Civil Eng. Mater. **7**(3) (2018b). https://doi.org/10.1520/acem20170103

Kaewunruen, S., Singh, M., Akono, A-T., Ishida, T.: Sustainable and self-sensing concrete. In: Proceedings of the 12th Annual Concrete Conference, Cha Am, Thailand (2017, invited). https://works.bepress.com/sakdirat_kaewunruen/80/

Manalo, A., Aravinthan, T., Karunasena, W., Stevens, N.: Analysis of a typical railway turnout sleeper system using grillage beam analogy. Finite Elements Anal. Des. **48**(1), 1376–1391 (2012)

Malveiro, J., Sousa, C., Riberiro, D., Calcada, R.: Impact of track irregularities and damping on the fatigue damage of a railway bridge deck slab. Struct. Infrastruct. Eng. (2018). https://doi.org/10.1080/15732479.2017.1418010

Meesit, R., Kaewunruen, S.: Vibration characteristics of micro-engineered crumb rubber concrete for railway sleeper applications. J. Adv. Concr. Technol. **15**(2), 55–66 (2017)

Neilsen, JCO.: Eigenfrequencies and eigenmodes of beam structures on an elastic foundation. J. Sound Vibr. **145**, 479–487 (1991)

Remennikov, A.M., Kaewunruen, S.: A review on loading conditions for railway track structures due to wheel and rail vertical interactions. Struct. Control Health Monit. **15**(2), 207–234 (2008)

Remennikov, A.M., Murray, M.H., Kaewunruen, S.: Reliability-based conversion of a structural design code for railway prestressed concrete sleepers. Proc. Inst. Mech. Eng. Part F J. Rail Rapid Transit **226**, 155–173 (2012)

Standards Australia: Australian Standard: AS1085.14-2003 Railway track material - Part 14: Prestressed concrete sleepers, Sydney, Australia (2003)

You, R., Li, D., Ngamkhanong, C., Janeliukstis, R., Kaewunruen, S.: Fatigue life assessment method for prestressed concrete sleepers. Front. Built Environ. **3**, 68 (2017). https://doi.org/10.3389/fbuil.2017.00068

Ülker-Kaustell, M., Karoumi, R.: Influence of non-linear stiffness and damping on the train-bridge resonance of a simply supported railway bridge. Eng. Struct. **41**, 350–355 (2012)

Vu, M., Kaewunruen, S., Attard, M.: Nonlinear 3D finite-element modeling for structural failure analysis of concrete sleepers/bearers at an urban turnout diamond. In: Handbook of Materials Failure Analysis with Case Studies from the Chemicals, Concrete and Power Industries, pp. 123–160. Elsevier, The Netherlands (2016). https://doi.org/10.1016/b978-0-08-100116-5.00006-5

Wolf, H.E., Edwards, J.R., Dersch, M.S., Barkan, C.P.L.: Flexural analysis of prestressed concrete monoblock sleepers for heavy-haul applications: methodologies and sensitivity to support conditions. In: Proceedings of the 11th International Heavy Haul Association Conference, Perth, Australia, June 2015

Zhai, W., Wang, S., Zhang, N., Gao, M., Xia, H., Cai, C., Zhao, C.: High-speed train–track–bridge dynamic interactions – part II: experimental validation and engineering application. Int. J. Rail Transp. **1**, 25–41 (2013)

Peridynamic Modeling of Rail Squats

Andris Freimanis[2], Sakdirat Kaewunruen[1(✉)], and Makoto Ishida[3]

[1] Riga Technical University, Riga, Latvia
s.kaewunruen@bham.ac.uk
[2] Railway and Civil Engineering, School of Engineering, The University of
Birmingham, Birmingham, UK
andris.freimanis_1@rtu.lv
[3] Railway Engineering Department, Nippon Koei Ltd., Tokyo, Japan
ishida-mk@n-koei.jp

Abstract. Rail squats and studs are typically classified as the propagation of any cracks that have grown longitudinally through the subsurface. Some of the cracks could propagate to the bottom of rails transversely, which have branched from the initial longitudinal cracks with a depression of rail surface. The rail defects are commonly referred to as 'squats' when they were initiated from damage layer caused by rolling contact fatigue, and as 'studs' when they were associated with white etching layer caused by the transform from pearlitic steel due to friction heat generated by wheel sliding or excessive traction. Such above-mentioned rail defects have been often observed in railway tracks catered for either light passenger or heavy freight traffics and for low, medium or high speed trains all over the world for over 60 years except some places such as sharp curves where large wear takes place under severe friction between wheel flange and rail gauge face. It becomes a much-more significant issue when the crack grows and sometimes flakes off the rail (by itself or by insufficient rail grinding), resulting in a rail surface irregularity. Such rail surface defect induces wheel/rail impact and large amplitude vibration of track structure and poor ride quality. In Australia, Europe and Japan, rail squats/studs have occasionally turned into broken rails. The root cause and preventive solution to this defect are still under investigation from the fracture mechanics and material sciences point of view. Some patterns of squat/stud development related to both of curve and tangent track geometries have been observed, and squat growth has also been monitored for individual squats using ultrasonic mapping techniques. This paper highlights peridynamic modeling of squat/stud distribution and its growth. Squat/stud growth has been measured in the field using the ultrasonic measurement device on a grid applied to the rail surface. The depths of crack paths at each grid node form a three dimensional contour of rail squat crack. The crack propagation of squats/studs is modelled using peridynamics. The modeling and field data is compared to evaluate the effectiveness of peridynamics in modelling rail squats.

© Springer Nature Switzerland AG 2019
S. El-Badawy and J. Valentin (Eds.): GeoMEast 2018, SUCI, pp. 108–118, 2019.
https://doi.org/10.1007/978-3-030-01911-2_10

1 Introduction

Rail squats, defined as cracks initiated from rolling contact fatigue (RCT) and from white etching layer (WEL); and growing longitudinally under the rail surface (in the direction to train), are a main problem for rail operators all around the globe. They are noticed by passengers when they create excess noise and vibration leading to uncomfortable rides (Remennikov and Kaewunruen 2008), but more importantly they can result in broken rail from the impact forces from wheel-rail interaction (Kaewunruen and Remennikov 2010; 2009). Additionally, squats have grown and turned into broken rails, which could result in a catastrophe (Ishida 2013). In practice, the rail surface defects have been a critical safety concern and key maintenance priority of railway infrastructure owners and managers who operate either low, moderate or high speed trains including passenger suburban, metro, urban, mixed-traffic and freight rail systems. The rail surface defects can cause high risks and significant consequences such as train derailments from rail breaks, component failures, and so on. It has been estimated that the cost of rail renewal program (rail replacement) due to rail squats and studs has become a significant portion of the whole track maintenance cost, reportedly in Australia, Asia, and European countries e.g. Austria, Japan, Germany and France (Kaewunruen et al. 2015; Kaewunruen and Ishida 2014; 2015; 2016). The rail squat/stud problem has largely been noticed when the ride quality of the passenger trains exceeds acceptable limits (Kaewunruen and Remennikov 2016). Excessive noise and vibration have later increased complaints against rail operators. Most importantly, the impact forces due to the wheel/rail interaction have undermined the structural integrity and stability of track components (Remennikov and Kaewunruen 2008). The study into innovative solutions for this problem is timely and significant. This paper is the world first to adopt peridynamic theory to predict dynamic crack growth from RCT, and to the best of authors' knowledge no other such work exists currently (Kaewunruen 2015; 2018).

   Classical mechanics theory uses spatial derivatives that do not exist in when the displacement field is discontinuous, so the techniques of fracture mechanics are used to study cracking phenomena. However, a major drawback is that crack path must be known *a priori*. Therefore, peridynamics (PD) (Silling 2000; Silling et al. 2007) was created to simulate objects with discontinuities. It uses integral not partial-differential equations, and deformation instead of strain to compute the internal forces. Since both are defined in the presence of cracks, PD are well suited for such analyses. Moreover, in PD crack path doesn't have to be known –cracks initiate automatically according to a prescribed damage law. These reasons make PD an excellent tool for studying different kinds of fracture and have been used to study damage in fiber-reinforced laminated composites (Colavito 2013; Hu et al. 2015; Hu et al. 2017; Kilic et al. 2009), glass (Bobaru et al. 2012; Bobaru and Zhang 2016), wood (Perré et al. 2015), concrete (Gerstle et al. 2009; Shen et al. 2013; Yaghoobi and Chorzepa 2015) and steel (De Meo et al. 2016).

2 State-Based Peridynamics

This section presents a brief overview of the state-based peridynamics theory, for an extended overview authors recommend (Bobaru et al. 2016; Madenci and Oterkus 2014; Silling and Lehoucq 2010). A peridynamic body consists of a number of nodes in the reference position x_i each describing some volume V_{x_i}. A node x_i interacts with other nodes x_j within a range called the horizon δ through bonds. Nodes within this range is called the family of x_i, H_{x_i}, When a body undergoes some deformation, node x_i experiences displacement u_i and moves to a deformed position y_i. This deformation creates a force density vector t_{ij} that depends on the collective deformation of H_{x_i} and t_{ji} that depends on the collective deformation of H_{x_j}. The bond deformation vectors and the force density vectors are stored in arrays called the deformation states, $\mathbf{Y}_{x_i}$ and the force states $\mathbf{T}_{x_i}$:

$$
\mathbf{Y}_{x_i} = \left\{ \begin{array}{c} y_1 - y_i \\ \vdots \\ y_n - y_i \end{array} \right\}, \mathbf{T}_{x_i} = \left\{ \begin{array}{c} t_{i1} \\ \vdots \\ t_{in} \end{array} \right\}.
\tag{1}
$$

It is common to write $\mathbf{T}(x_i)\langle x_j - x_i \rangle$ when referring to a force density vector t_{ij} in a specific bond $\boldsymbol{\xi}_{ij} = x_j - x_i$. The force density vectors depend on the deformation, so we can write:

$$
\mathbf{T}(x_i) = \mathbf{T}(\mathbf{Y}(x_i))
\tag{2}
$$

where the function $\mathbf{T}(x_i)$ is the material model. Then the peridynamic equation of motion in the integral form is

$$
\rho(x_i)\ddot{u}(x_i, t) = \int_{H_{x_i}} (\mathbf{T}(x_i)\langle x_i - x_j \rangle - \mathbf{T}(x_j)\langle x_j - x_i \rangle) dV_{xj} + b(x_i)
\tag{3}
$$

where $\rho(x_i)$ – density, $\ddot{u}(x_i)$ – acceleration and $b(x_i)$ – external force density.

Boundary conditions are not required in PD solution, because the PD equation of motion does not contain any spatial derivatives, however, they are necessary to solve many real-life problems. Since nodes describe some volume, boundary conditions must also be applied to some volume.

Damage is introduced by breaking a bond. The simplest damage criterion could be the critical stretch, in which a bond breaks when it's stretched past some critical value s_c:

$$
\omega(x_i) = \left\{ \begin{array}{ll} 1, & \text{if } s_{ij} < s_c \\ 0, & \text{if } s_{ij} \geq s_c \end{array} \right., \; s_{ij} = \frac{|y_j - y_i| - |x_j - x_i|}{|x_j - x_i|} = \frac{|\mathbf{Y}\langle \boldsymbol{\xi}_{ij} \rangle| - |\boldsymbol{\xi}_{ij}|}{|\boldsymbol{\xi}_{ij}|}.
\tag{4}
$$

The damage at a node is defined in (Silling and Askari 2005) as a ratio between the broken and the initial number of bonds:

$$\phi(x_i) = 1 - \frac{\int_{H_{x_i}} \omega(x_i) dV_{\xi_{ij}}}{\int_{H_{x_i}} dV_{\xi_{ij}}}. \tag{5}$$

The PD fatigue damage model used in this study was introduced in (Silling and Askari 2014) and used in (Jung and Seok 2017; 2016; Zhang et al. 2016). Other researchers have also developed fatigue damage models (Baber and Guven 2017; Oterkus et al. 2010), however, these models use bond-based PD and simulate only the crack growth phase. The overview of the model is given in (Silling and Askari 2014), but it's repeated here for completeness.

A body undergoes some cyclic deformation between two extremes + and -, bond strains at each extreme and the cyclic bond strain is:

$$s_{ij}^+ = \frac{|\mathbf{Y}^+ \langle \xi_{ij} \rangle| - |\xi_{ij}|}{|\xi_{ij}|}, \quad s_{ij}^- = \frac{|\mathbf{Y}^- \langle \xi_{ij} \rangle| - |\xi_{ij}|}{|\xi_{ij}|}, \quad \varepsilon_{ij} = |s_{ij}^+ - s_{ij}^-|. \tag{6}$$

For each bond a variable called the "remaining life" $\lambda_{ij}(x_i, \xi_{ij}, N)$ is defined. It degrades at each loading cycle N, and a bond breaks when the remaining life is reduced to zero:

$$\lambda_{ij}(N) \leq 0. \tag{7}$$

At the beginning when $N = 0$:

$$\lambda_{ij}(0) = 1, \tag{8}$$

then at each cycle in crack nucleation phase (phase I) the change of λ is given by

$$\frac{d\lambda_{ij}}{dN}(N) = \begin{cases} -A_1 (\varepsilon_{ij} - \varepsilon_\infty)^{m_1} & , \text{if } \varepsilon_{ij} > \varepsilon_\infty \\ 0 & , \text{if } \varepsilon_{ij} \leq \varepsilon_\infty \end{cases}, \tag{9}$$

where ε_∞ - fatigue limit under which no fatigue damage occurs, A_1, m_1 – parameters for phase I. In the phase II the remaining life changes according to:

$$\frac{d\lambda_{ij}}{dN}(N) = -A_{11} \varepsilon_{ij}^{m_{11}}. \tag{10}$$

The transition from phase I to phase II is handled by applying the phase I model with parameters A_1, m_1 to a node x_i till there is a node x_j in H_{x_i} with damage

$$\phi(x_j) \geq 0.5, \tag{11}$$

then reset the remaining life of the bonds connected to x_i to 1 and switch to phase II model.

3 Computational Model

The fatigue damage model was implemented in the open-source Peridynamics program Peridigm (Lihlewood et al. 2013; Parks et al. 2012). For quasi-static analysis the acceleration is zero, so peridynamic equation of motion is approximated as:

$$\sum_{H_{x_i}} (T[x_i, t]\langle x_j - x_i \rangle - T[x_j, t]\langle x_i - x_j \rangle)\Delta V_{x_j} + b(x_i, t) = 0. \tag{12}$$

and then solved using Newton's method. The remaining life of each bond is computed after each simulation step, by:

$$\lambda_{i,j}^0 = 1, \quad \lambda_{i,j}^n = \lambda_{i,j}^{n-1} - \Delta N A(\varepsilon_{i,j}^n)^m. \tag{13}$$

(Silling and Askari 2014) introduces two techniques – implicit strain simulation and time mapping – to speed up simulations. Both were used and are repeated here for completeness. In case of a high-cycle fatigue the bond strains from the train wheel load are below the elastic limit, so an elastic material model is used to simulate the rail. In such a case strain in a bond would change linearly between + and – loading conditions, so it is possible to simulate only the + loading condition and compute

$$s^- = Rs^+, \quad R = \frac{P^-}{P^+}, \tag{14}$$

where R – loading ratio, P – applied load at each extreme. The cyclic strain is then given by:

$$\varepsilon = |s^+ - s^-| = |(1 - R)s^+|. \tag{15}$$

Simulation time is mapped against the current cycle using a linear mapping:

$$N = \frac{t}{\tau}, \tag{16}$$

Where τ is a constant. Then remaining life change in current simulation time is found through:

$$\frac{\Delta\lambda}{\Delta t} = \frac{\Delta\lambda}{\Delta N}\frac{\Delta N}{\Delta t}. \tag{17}$$

4 Model of a Rail

The model was discretized using meshless method introduced in (Silling and Askari 2005). Each node has a position in 3D space and describes some amount of volume around it. The edge length of a node was 0.001 m and the horizon $\delta = 0.00201$ m. The mesh is rather coarse and the horizon rather short, these values were selected to reduce computational expense, because here the preliminary results of a larger work are presented. In this study a model of UIC60 profile rail head was used. Due to irregular form of the rail head, nodes were not perfectly cubic; however, differences were insignificant near the middle of the rail head and increased only slightly near the gauge corners. The shape of the rail head was first created with solid hexahedral elements in finite element (FE) program Ansys, then element centroids and volumes were exported to a text file to be used as a mesh for Peridigm. Movement in vertical (y) direction was fixed for a layer of nodes with thickness δ at the bottom of the rail head, additionally damage was forbidden for nodes less than 3δ from bottom, to avoid unphysical behavior near the boundary conditions.

Since applied loads didn't cause the material to exceed its yield strength, an elastic material model was used, it's properties: density – 7850 kg/m^3, Poisson's ratio – 0.3, Young's modulus – 210 GPa. The instructions on how to obtain damage model parameter values are included in (Silling and Askari 2014), they were applied to test data presented in (Scutti et al. 1984). While the data were quite old, they were used, because all four fatigue model parameters could be estimated from just one data source, since both E-N curves and Paris law plots are presented. Fatigue damage model parameters were: A_1–426.00, m_1–2.77, A_2–3249.00, m_2–4.00. The model switched from phase I to phase II, when damage at a node reached 40%. Also, a linear time mapping was used, with τ equal to 0.001 till 21'000 cycles, 0.01 till 26'200 cycles, and 0.1 till 26670th cycle.

5   Train Wheel Loading

Two different train wheel loadings were applied – vertical pressure and surface shear traction. They were obtained from (Wei et al. 2016) Fig. 5f and 6f, which show elastic pressure and elastic surface shear traction respectively. Loads were applied to a single layer of nodes at the middle of rail head's top surface, with a loading ratio $R = 0$. The wheel-rail contact area (Fig. 4f in (Wei et al. 2016)) was approximated with an ellipse whose half-axis were a = 0.0066 m, c = 0.006386 m. The elastic pressure was approximated by an ellipsoid with half-axis a = 0.0066 m, c = 0.006386 m, b = 1.116GPa, then the value at each node was calculated from ellipsoid formula and converted to force density. Shear traction forces change based on the location of a node in the contact area, so they were described by four tri-linear functions (Q1 to Q4). First, nodes were split into seven intervals based on their transversal location (z coordinate), second a tri-linear shear traction function was selected based on the interval, finally shear traction value was calculated and converted to force density. Table 1 shows interval limits and corresponding surface shear traction functions, Fig. 1 presents the surface shear traction functions.

Table 1. Interval limits in transverse direction and their tri-linear surface shear traction functions.

Interval	Interval start transversal coordinate, m	Interval end transversal coordinate, m	Tri-linear Surface Shear Traction Function
1	–0.006386	–0.005	Q4
2	–0.005	–0.003	Q3
3	–0.003	–0.001	Q2
4	–0.001	0.001	Q1
5	0.001	0.003	Q2
6	0.003	0.005	Q3
7	0.005	0.006386	Q4

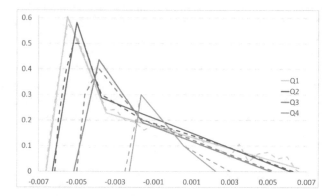

Fig. 1. Surface shear traction (GPa) vs the x coordinate of a node (m). Dashed lines show data from (Wei et al. 2016) and solid lines show the fitted tri-linear functions (Q1 to Q4).

6 Results

Simulation ran for 26'670 cycles. Figure 2 shows damage in the rail heads cross-section parallel to the longitudinal direction. First bonds break before 14'000th cycle, then damage zone under applied loads expands till at 22'600th cycle node reaches 40% damage (not showed in Fig. 2), and its neighborhood gets switched to phase II. Damage develops quite rapidly after that, leading to near complete damage (80%) at a node at cycle 26'670. Most of the broken bonds are concentrated under the maximum pressure as expected, because it creates larger bond strains. This simulation shows a very fast onset of damage that doesn't grow into longer cracks, instead damage is contained under the load area. However, it can be seen in Fig. 2 that damage extends forward under the surface forming the beginning of a squat. Longer simulations are needed to see if damage really expands in that direction. Figure 3 shows damage in cross-section perpendicular to the longitudinal direction. The damage is nearly sym-metrical at 20'000 cycles, but asymmetrical at 26'670 cycles. Most likely this is caused by the iterative solver converging to slightly different values and the mesh not being completely symmetric.

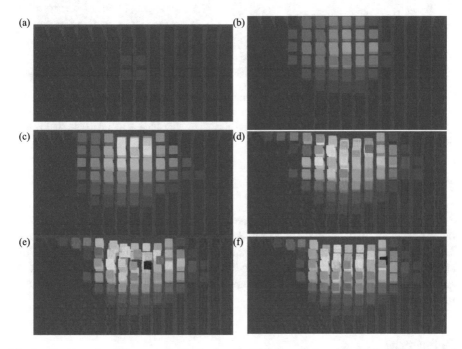

Fig. 2. Cross-section (plane parallel to the longitudinal direction) at the middle of the rail head showing damage from rolling contact fatigue at (a) - 14'000 cycles, (b) – 21'000, (c) – 25'000, (d) – 26'200, (e) – 26'580, (f) – 26'670. Rolling direction is to the right and only damaged zone is shown. Blue color – less damage, red – more damage.

Additionally, Fig. 3(b) shows that the top layer of nodes has broken off the following layers and penetrated them, which is an unphysical behavior. Since a large (> 80%) number of bonds are broken for the top layer and loads are applied only to this layer, they no longer are distributed further down. This is the likeliest explanation for why cracks do not grow into longer squats. This problem could be remedied with finer mesh, because the same loads will be distributed over larger number of nodes, resulting in less deformation at each node and, therefore, bond. Applying loads to a thicker layer could help distribute the wheel loading in way that's closer to real life conditions. It was suggested in (Macek and Silling 2007) that loads are applied to a layer equal to the horizon. Finally, selecting longer horizon could help distribute loading to lower layers.

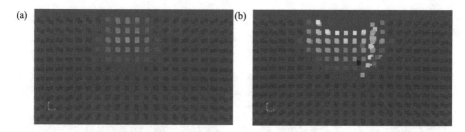

Fig. 3. Cross-section (plane normal to longitudinal direction) at the middle of rail head (center of the applied load) showing damage from rolling contact fatigue at (a) - 20'000 cycles, (b) – 26'670. Color scheme indicates damage at a node with blue – less damage, red – more damage.

7   Conclusions

Peridynamic theory can be used to simulate rolling contact fatigue damage in a railway rail. The maximum damage will occur under the maximum pressure values as expected. The simulation showed onset of a rail squat, however simulations limitations didn't allow author's to further explore it. To remedy this boundary conditions must be applied in a way that ensures continuous load transfer to lower layers, even after surface layer sustains large damage.

References

Baber, F., Guven, I.: Solder joint fatigue life prediction using peridynamic approach. Microelectron. Reliab. **79**, 20–31 (2017). Accessed https://doi.org/10.1016/j.microrel.2017.10.004

Bobaru, F., Foster, J.T., Geubelle, P.H., Silling, S.A.: Handbook of Peridynamic Modeling. CRC Press, Boca Raton (2016)

Bobaru, F., Ha, Y.D., Hu, W.: Damage progression from impact in layered glass modeled with peridynamics. Cent. Eur. J. Eng. **2**(4), 551–561 (2012)

Bobaru, F., Zhang, G.: Why do cracks branch? a peridynamic investigation of dynamic brittle fracture. Int. J. Fract. (2016). Accessed http://link.springer.com/10.1007/s10704-015-0056-8

Colavito, K.: Peridynamics for Failure and Residual Strength Prediction of Fiber-Reinforced Composites (2013). Accessed http://search.proquest.com/docview/1475241115?accountid=45156

Gerstle, W.H., Sau, N., Sakhavand, N.: On peridynamic computational simulation of concrete structures. ACI Special Publication (February 2016) (2009). Accessed http://www.di.uson.mx/departamentos/dicym/images/doc/SP_265_11.pdf

Hu, Y.L.L., De Carvalho, N.V.V., Madenci, E.: Peridynamic modeling of delamination growth in composite laminates. Compos. Struct. **132**, 610–620 (2015). Accessed http://linkinghub.elsevier.com/retrieve/pii/S0263822315004626

Hu, Y., Madenci, E., Phan, N.: Peridynamics for Predicting Damage and Its Growth in Composites, pp. 1214–1226 (2017)

Ishida, Makoto: Rolling contact fatigue (RCF) defects of rails in Japanese railways and its mitigation strategies. Electron. J. Struct. Eng. **13**(1), 67–74 (2013)

Jung, J., Seok, J.: Fatigue crack growth analysis in layered heterogeneous material systems using peridynamic approach. Compos. Struct. **152**, 403–407 (2016). Accessed http://dx.doi.org/10.1016/j.compstruct.2016.05.077

Jung, J., Seok, J.: Mixed-mode fatigue crack growth analysis using peridynamic approach. Int. J. Fatigue **103,** 591–603 (2017). Accessed http://dx.doi.org/10.1016/j.ijfatigue.2017.06.008

Kaewunruen, S.: Identification and prioritization of rail squat defects in the field using rail magnetisation technology. Proc. SPIE – Int. Soc. Opt. Eng. **9437**, 94371H (2015). https://doi.org/10.1117/12.2083851

Kaewunruen, S.: Discussion of "field test performance of noncontact ultrasonic rail inspection system" by Stefano Mariani, Thompson Nguyen, Xuan Zhu, and Francesco Lanza di Scalea. J. Transp. Eng. Part A Syst. **144**(4), 07018001 (2018). https://doi.org/10.1061/JTEPBS.0000134

Kaewunruen, S., Ishida, M.: Field monitoring of rail squats using 3D ultrasonic mapping technique. J. Can. Inst. Non-destr. Eval. (invited) **35**(6), 5–11 (2014)

Kaewunruen, S., Ishida, M.: Rail squats: understand its causes, severity, and non-destructive evaluation techniques. In: Proceedings of the 20th National Convention on Civil Engineering, Pattaya, Thailand, 8–10 July 2015. (Best Paper Award in Infrastructure Engineering), https://works.bepress.com/sakdirat_kaewunruen/56/

Kaewunruen, S., Ishida, M.: In situ monitoring of rail squats in three dimensions using ultrasonic technique. Exp. Tech. **40**(4), 1179–1185 (2016)

Kaewunruen, S., Ishida, M., Marich, S.: Dynamic wheel-rail interaction over rail squat defects. Acoust. Aust. **43**(1), 97–107 (2015). https://doi.org/10.1007/s40857-014-0001-4

Kaewunruen, S., Remennikov, A.M.: Dynamic properties of railway track and its components: recent findings and future research direction. Insight - Non-Destruct. Test. Cond. Monit. **52**(1), 20–22 (2010). Accessed http://www.ingentaconnect.com/content/bindt/insight/2010/00000052/00000001/art00006

Kaewunruen, S., Remennikov, A.M.: Progressive failure of prestressed concrete sleepers under multiple high-intensity impact loads. Eng. Struct. **31**(10), 2460–2473 (2009). Accessed http://dx.doi.org/10.1016/j.engstruct.2009.06.002

Kaewunruen, S., Remennikov, A.M.: Current state of practice in railway track vibration isolation: an Australian overview. Aust. J. Civil Eng. **14**(1), 63–71 (2016)

Kilic, B., Agwai, A., Madenci, E.: Peridynamic theory for progressive damage prediction in center-cracked composite laminates. Compos. Struct. **90**(2), 141–151 (2009). Accessed http://dx.doi.org/10.1016/j.compstruct.2009.02.015

Lihlewood, D., Parks, M., Mitchell, J., Silling, S.: The Peridigm Framework for Peridynamic Simulations, July 2013

Macek, R.W., Silling, S.A.: Peridynamics via finite element analysis. Finite Elem. Anal. Des. **43**(15), 1169–1178 (2007)

Madenci, E., Oterkus, E.: Peridynamic Theory and Its Applications. Springer, New York (2014). Accessed http://link.springer.com/book/10.1007/978-1-4614-8465-3

De Meo, D., Diyaroglu, C., Zhu, N., Oterkus, E., Siddiq, M.A.: Modelling of stress-corrosion cracking by using peridynamics. Int. J. Hydrog. Energy **41**(15), 6593–6609 (2016). Accessed http://dx.doi.org/10.1016/j.ijhydene.2016.02.154

Oterkus, E., Guven, I., Madenci, E.: Fatigue failure model with peridynamic theory. In: 2010 12th IEEE Intersociety Conference on Thermal and Thermomechanical Phenomena in Electronic Systems, ITherm 2010 (2010)

Parks, M.L., Littlewood, D.J., Mitchell, J.A., Silling, S.A.: Peridigm Users' Guide (2012)

Perré, P., Almeida, G., Ayouz, M., Frank, X.: New modelling approaches to predict wood properties from its cellular structure: image-based representation and meshless methods. Ann. Forest Sci. (2015). http://link.springer.com/10.1007/s13595-015-0519-0

Remennikov, A.M., Kaewunruen, S.: A review of loading conditions for railway track structures due to train and track vertical interaction. Struct. Control Health Monit. **15**(2), 207–234 (2008). Accessed https://doi.org/10.1002/stc.227%0A

Scutti, J.J., Pelloux, R.M., Fuquen-Moleno, R.: Fatigue behavior of a rail steel. Fatigue Fract. Eng. Mater. Struct. **7**(2), 121–135 (1984). Accessed http://doi.wiley.com/10.1111/j.1460-2695.1984.tb00410

Shen, F., Zhang, Q., Huang, D.: Damage and Failure Process of Concrete Structure under Uniaxial Compression Based on Peridynamics Modeling (2013)

Silling, S.A.: Reformulation of elasticity theory for discontinuities and long-range forces. J. Mech. Phys. Solids **48**(1), 175–209 (2000)

Silling, S.A., Askari, E.: A meshfree method based on the peridynamic model of solid mechanics. Comput. Struct. **83**(17–18), 1526–1535 (2005)

Silling, S.A., Epton, M., Weckner, O., Xu, J., Askari, E.: Peridynamic states and constitutive modeling. J. Elast. **88**(2), 151–184 (2007)

Silling, S.A., Lehoucq, R.B.: Peridynamic theory of solid mechanics. Adv. Appl. Mech. **44**, 73–168 (2010)

Silling, S., Askari, A.: Peridynamic Model for Fatigue Cracks. SANDIA REPORT SAND2014-18590, Albuquerque (2014). http://docs.lib.purdue.edu/ses2014/mss/cfm/22/

Wei, Z., Li, Z., Qian, Z., Chen, R., Dollevoet, R.: 3D FE modelling and validation of frictional contact with partial slip in compression–shift–rolling evolution. Int. J. Rail Transp. **4**(1), 20–36 (2016). Accessed http://dx.doi.org/10.1080/23248378.2015.1094753

Yaghoobi, A., Chorzepa, M.G.: Meshless modeling framework for fiber reinforced concrete structures. Comput. Struct. **161**, 43–54 (2015). Accessed http://www.sciencedirect.com/science/article/pii/S0045794915002485

Zhang, G., Le, Q., Loghin, A., Subramaniyan, A., Bobaru, F.: Validation of a peridynamic model for fatigue cracking. Eng. Fract. Mech. **162**, 76–94 (2016). Accessed http://dx.doi.org/10.1016/j.engfracmech.2016.05.008

Influence of Oil Palm Shell (OPS) on the Compaction Behavior and Strength Improvement of Soil-OPS Composites: A Pilot Study

Shi Jun Loi[1], Mavinakere Eshwaraiah Raghunandan[1,3(✉)],
Susilawati[1], and Tan Boon Thong[2]

[1] Civil Engineering, School of Engineering, Monash University,
Bandar Sunway, Malaysia
mavinakere.raghunandan@monash.edu
[2] Mechanical Engineering, School of Engineering, Monash University,
Bandar Sunway, Malaysia
[3] Advanced Engineering Platform, Monash University,
Bandar Sunway, Malaysia

Abstract. In construction industry, state-of-the-art experimental trials such as plastic road by Scottish and rubber road in Malaysia have shown potential value in the utilization of locally available wastes in engineering projects. Malaysia, the second largest global producers of palm oil and related products, estimates millions of tons of biomass by-product annually. Oil Palm Shell (OPS) is one amongst the biomass thus produced. OPS is well noted in literatures for its low specific gravity and better strength/stiffness, though not in absolute comparison to aggregates. Thus, an eco-composite with OPS supplementing conventional sub-base material hypothetically is a promising alternative, facilitating productive utilization of OPS – an agricultural waste, reducing the self-weight of the sub-base, and soil reinforcement leading to improved strength. The main aim of this paper is to study the influence of OPS content on the compaction behavior and strength improvement of soil-OPS composites for its application as sub-base material in rural roads. Basic properties of the OPS samples and three soil samples locally procured – kaolin, medium sand and clayey sand, were evaluated in laboratory. The performance of soil-OPS composite in terms of their compaction and California bearing ratios is evaluated and discussed in this paper. In conclusion, the OPS was observed to have significant influence on the compaction characteristics and strength (CBR). In brief, the porous surface texture of OPS is most likely the key parameter controlling the compaction parameters, while contributions from the tensile strength of individual OPS samples potentially has a significant influence on the strength (CBR) improvements observed in this pilot study.

1 Introduction

Owing to growing global environmental concerns and entrenched appetite to enforce green technology and practices, eco-composites have gained strong position as alternative in various industries including construction industry. Efforts to produce alternative

© Springer Nature Switzerland AG 2019
S. El-Badawy and J. Valentin (Eds.): GeoMEast 2018, SUCI, pp. 119–132, 2019.
https://doi.org/10.1007/978-3-030-01911-2_11

materials, particularly using various by-products and/or wastes is one such step forward to address these potential environmental concerns. For example, in Malaysia, collaborative studies between the Malaysia rubber board and the public works department have exercised suitable efforts to evaluate the performance of Cuplump Modified Asphalt (CMA) for pavement construction. Positive results from the CMA treatment, further led its implementation on a 10 km long rural road in Malaysia (The Sun Daily 2017). An attempt to utilize recycled waste plastic bottle for road surfacing by the Scottish is perhaps noteworthy as well. This study showed the bitumen substituent (MR6) used as a filler material was 60% stronger and could serve about 10 times longer than the conventional asphalt (Jeffay 2018). Not limiting to these examples alone, literature further demonstrates utilization of various such waste materials including low density polyethylene and crumb rubber (Rokade 2012), natural rubber (Ibrahim 2018) to name some as viable substitutes in highway pavements and construction practices. Nonetheless, Malaysian palm biomasses also demonstrates immense potential in the process of subgrade stabilization in pavements and can be hypothesized as another viable by-product be used in rural roads and/or highways.

Malaysia, one of the world's top producers of palm oil and its derivatives, estimates to around 19.8 million tons of palm oil supply globally in 2015 (MPOB 2016). Moreover, the domestic palm oil production is forecasted to increase to around 25 million tons in 2035 (Yean and ZhiDong 2012). Undeniable, the yield of palm oil industry has marked as one of the most significant contributors to Malaysia's economy. However, a lot of biomass and/or wastes are likely produced during the palm oil production, namely oil palm shell (OPS), oil palm trunks, oil palm fronds, empty fruit bunches, palm pressed fibers, and palm oil mill effluent (Loh 2017; Shuit et al. 2009). These by-products, owing to their abundance, reportedly create major disposal and related problems (Abdullah and Sulaiman 2013). More emphasis has always been on improper planning and disposal leading to environmental concerns. To this end, OPS likely participates in this category when speaking about its disposal – which in current practice is either incinerated to produce heat, used as surface covers in plantation pathways or roads, and sometime landfilled (Abdullah and Sulaiman 2013; Loh 2017; Mohammed et al. 2011; Sumathi et al. 2008). Notwithstanding the current practice perhaps, to this end application of OPS as waste-to-wealth alternative in the stabilization soil subgrades for rural roads is emphasized.

OPS, also known as palm kernel shell are generally listed amongst the fibrous by-products produced during the process of palm oil extraction. Sharma et al. (2015) defined OPS as the fractions left after removing the crushed nut in the oil palm mill, stockpiled and stored at open space following the separation process. The properties of OPS such as density are governed by the age and species of OPS (Yew et al. 2014). The low specific gravity (G_S) defined as the ratio of density of OPS to that of water, is generally between 1.14 and 1.62 – often less than 2. This low G_S value perhaps is one of the main reasons for OPS to locate its place replacing the aggregate in conventional concrete aiding to the production of lightweight concrete (Okpala 1990). The low GS values also explain the low bulk density (ρ_b) of the OPS which is reported to range from 500 kg/m^3 to 600 kg/m^3, and the literature suggests an inversely proportional relation between the ρ_b and the size of OPS (Alengaram et al. 2010). Water absorption is another essential property which governs various behavior and properties of organic matters. The high-water absorption of such organic materials is attributed to the present of pores on the surface of the shell (Alengaram et al. 2010; Alengaram et al. 2013). Water absorption of OPS is highly influenced by the species of the palm tree (Yew et al. 2014), with the 24-h water absorption of OPS closely ranging between 14 and 27% (Alengaram et al. 2013).

In summary, OPS is well noted in the literature for its low specific gravity in comparison to other conventional construction materials, which makes it a promising substituent replacing coarse and fine aggregates in concrete to produce lightweight concrete. Literature also highlight the strength/stiffness of OPS, though not in absolute comparison to aggregates, are in good comparison when the compressive strength of OPS-concrete compared with other light-weight concretes (Alengaram et al. 2013). The hemispherical dome shape of the OPS – potentially enabling better load distribution, perhaps is one of a key parameter which is a crucial requirement in the road sub-grade as well. Thus, an eco-composite utilizing OPS supplemented to the conventional sub-grade material hypothetically is a promising alternative, facilitating productive utilization of OPS – an agricultural waste, reducing the self-weight of the sub-base, and soil reinforcement leading to improved strength. The main aim of this paper therefore to address this possibility, supported by discussions using data available in the literature and results obtained from selected laboratory tests conducted on OPS-supplemented-soil samples. Three different soil samples including commercially procured kaolin sample, sand and clayey sand sampled from locally from Gerik town in the state of Perak at the Peninsular Malaysia.

2 Overview of the Current Application of Ops

Huge quantities of the OPS generated annually has found its aspiration to be utilized as renewable energy – a state of the art research path and one of the mains focuses in the 11th Malaysian Plan (Economic Planning Unit 2015). Literature highlights utilization of oil palm biomass such as empty fruit bunch, oil palm fiber, trunks, fronds and OPS to produce hydrogen, which eventually used as synthetic fuel with high energy efficiencies and lesser secondary waste and environmental impacts (Awalludin et al. 2015). OPS is often incinerated to generate steam and electricity within the plantations or mills to support certain manufacturing processes (Shuit et al. 2009). The presence of function groups such as carboxylic and hydroxyl in OPS, perhaps has qualified it as a suitable material to remove heavy metal especially chromium from industrial wastewater (Nomanbhay and Palanisamy 2005). In addition, acid treated OPS charcoal – an OPS derivative of OPS, has been utilized in wastewater treatment (Nomanbhay and Palanisamy 2005). While another study reveals the application of OPS as material filler and sorbent for water treatment (Okoroigwe et al. 2014).

Various studies in literature ascertain the suitability and potential usage of OPS replacing and/or partially substituting conventional aggregates (i.e. gravel, sand) in concrete to produce lightweight concrete (Mannan and Ganapathy 2001, 2002; Ndoke 2006; Okafor 1988; Olanipekun et al. 2006; Teo et al. 2007). Discussion in these literatures suggest that the OPS concrete is in good comparison with the performance of other lightweight concretes. For example, an attempt by Mannan and Ganapathy (2001) to produce lightweight concrete using OPS and fly ash showed that the OPS-fly ash concrete to achieve 28-day compressive strength of 24.20 N/mm^2. In general, the overall strength of OPS concrete is significantly influenced by the strength of individual OPS (Okafor 1988). On the down side however, the compressive strength of OPS concrete is inversely proportional to the OPS content in the concrete (Olanipekun et al. 2006). Catering to this finding, literature also present optimized OPS contents in concrete, for examples Ndoke (2006) suggested a limit of 30% OPS to replace coarse aggregate on concretes. Nonetheless, recent attempt to use OPS as aggregates has to produce high strength lightweight concrete by (Shafigh et al. 2011a, b) is not be ignored. Authors reported high workability and improved compressive strengths using crushed OPS in concrete, moreover they compared the OPS concrete to result in strength equivalent to a grade M30 conventional concrete. Not limiting to OPS concretes alone, attempts to utilize OPS in concrete pavement to support lightly traffic loads are also noted in the literature (Khankhaje et al. 2017).

Soil stabilization predominantly referring to strengthening and/or increasing the stability – increased soil bearing capacity, reducing compressibility, and improving drainage, is emphasized in this paper. Thus, existing literature focusing of the utilization of OPS for this purpose, though scarce, has been reviewed henceforth. An attempt by Gungat et al. (2013) showed an increasing trend in the load bearing capacity of clay soil with increasing OPS content based on the laboratory California Bearing Ratio test results, while the load bearing capacity was reported to deteriorate with time (or curing). Another recent research by Adeboje et al. (2017) reported the utilization of OPS in stabilizing lateritic soils. The experiment results surprisingly revealed the presence of OPS in lateritic soil to decrease the optimum moisture content and increase the maximum dry density. Though the unconfined compressive strength (UCS) increased with the addition of OPS in laterites, it is apparent that OPS alone is insufficient to treat lateritic soils (Adeboje et al. 2017; Ekeocha and Agwuncha 2014). On the other hand, the biodegradability and durability of the natural materials like oil palm shell and fibers have always posed an immense concern limiting the ready application on OPS construction industry, particularly in stabilizing soils. However, it is worth mentioning that attempts to address this issue by applying chemical coatings on fibers using polymer compounds and others are presented as well (Prabakar and Sridhar 2002), but are yet in the primitive stages of research.

3   Material and Methodology

Raw OPS samples in this study were procured from a mill/plantation located in the Klang region of the Selangor state at Peninsular Malaysia. As anticipated, the raw OPS samples transported to the laboratory from field, was observed to contain different moisture contents with different batches. Thus, these OPS samples were pre-processed to remove moisture using oven drying method in the laboratory – under a constant temperature of 105 °C for 24 h prior to other experiments. Laboratory sieve analysis was then conducted to map the particle size distribution of the OPS samples using procedure similar to specifications in ASTM D6913/D6913 M - 17. Figure 1 shows the particle size distribution curve for the OPS, which suggest a poorly graded distribution with more than 90 per cent of the particles retained on 2 mm opening sieves amongst the particles that were finer than 20 mm effective diameter. Basically, most of the OPS samples fall within a very narrow range of particle sizes. However, owing to the dimensions of the proposed samples, the boundary conditions and requirements, and the purpose of OPS's presence (reinforcement) in the soil composites, OPS samples with particle sizes between 4.75 mm and 15 mm only was selected and used for testing in this study. Table 1 tabulates and the other key physical and initial properties of the OPS samples used in this study.

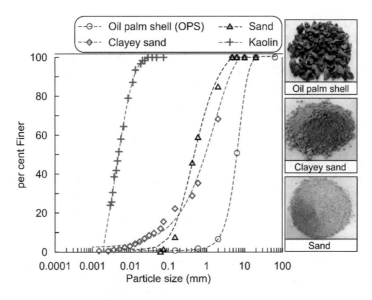

Fig. 1. Particle size distribution curve of the representative sample of OPS

Three soil samples including a kaolin, sand and clayey sand samples were used to host the OPS. Figure 1 also shows the particle size distributions (PSD) curves for all three soil samples. The PDS or gradation tests in this study were conducted as per the specifications of ASTM D6913-17. It should be note that the PSD data points for the kaolin sample alone was referred from Sriraam (2018). The kaolin sample was pre-processed and cleaned sample commercially procured, while the sand sample was prepared to a desired gradation in the laboratory from a heap of river sand sample. The clayey sand samples were sampled from a construction site in the Pulau Banding region of the Gerik town in the state of Perak at the Peninsular Malaysia. Table 2 highlights details of the key geotechnical properties and the soil classification for all three soil samples, and the procedure followed in the process. Four different soil-OPS composites were then prepared using 5, 10, 20, and 30 per cent by weight of OPS samples with each soil samples, in addition to one control sample – with no OPS added. Nonetheless, owing to significant reduction in the GS value of OPS win comparison to the soil, OPS content is also expressed as per cent by volume of soil in this paper. This is reflected in the tables within the parenthesis where the OPS content is expressed. However, OPS content in the figures are still retained as per cent by weight of soil for simplicity. The

Table 1. Properties of OPS

Soil types	OPS
Maximum grain size (mm)	15.3
Shell thickness (mm)	0.3–3.0
Field moisture content (%)	13.70–24.50
Organic content (%)	72.50–87.80
24-h water absorption (%)	33.18 – 39.00

samples were prepared using repeated hand mixing of dry OPS and soil samples, with sue care emphasized on their storage in the laboratory prior to testing. Thus, a total of 15 (3 soil samples × 5 OPS contents) variations were obtained and ready for testing.

Table 2. Geotechnical properties of targeted soils

Soil types	Kaolin	Sand	Clayey sand	Procedure
Specific gravity	2.46	2.68	2.65	ASTM D854-13
Liquid limit (LL)	73.64	–	39.83	ASTM D4318-05
Plastic limit (PL)	37.86	–	22.65	ASTM D4318-05
Plasticity index (PI)	35.77	–	17.18	ASTM D4318-05
Soil classification (USCS)	MH	SW	SC	ASTM D2487-00

The optimum moisture content (w_{opt}) and the maximum dry unit weight ($\gamma_{d,max}$) values of the tests samples were determined in the laboratory using the standard compaction test in accordance with ASTM D698-12. While the California Bearing Ration (CBR) test was conducted (as per the BS1377-4:1990) to ascertain the performance of the proposed soil-OPS composite's utilization in the sub-grade layers of rural roads based on their penetration resistance values (Gray and Al-Refeai 1986; Kumar et al. 1999; Li et al. 2014; Yetimoglu and Salbas 2003). The test samples were initially prepared to their respective w_{opt} and $\gamma_{d,max}$ determined from the compaction tests. The sample were compacted in 3 layers with 62 blows for each layer in the standard CBR mold.

To facilitate a detailed study on the performance of OPS-soil composites under different possible conditions, soaked and un-soaked test samples were prepared for the CBR test. In addition, a 4.5 kg surcharge was applied on the top of test specimen prior and during the testing to simulate traffic weight on the road surface as per the procedure. The soaked samples were immersed in the water, along with the surcharge, for 96 h prior to conducting the CBR test. Due care was exercised measuring any swelling or displacement in test sample every 24 h, to carefully monitor sample's behavior under soaked conditions over time.

4   Results and Discussion

4.1   Effect of OPS Content on Compaction Properties

Compaction tests in this study were mainly aligned to determine the w_{opt} and $\gamma_{d,max}$ values of the OPS-soil composite proposed in this study, the effect of increasing OPS content of the w_{opt} and $\gamma_{d,max}$ values are also discussed. Figure 2 shows the outcomes obtained from the standards proctor compaction tests conducted on all 3 soils with varying content of OPS in this study. Tables 3, 4 and 5 further tabulate the w_{opt} and $\gamma_{d,max}$ values obtained for each variants of the OPS-soil composites.

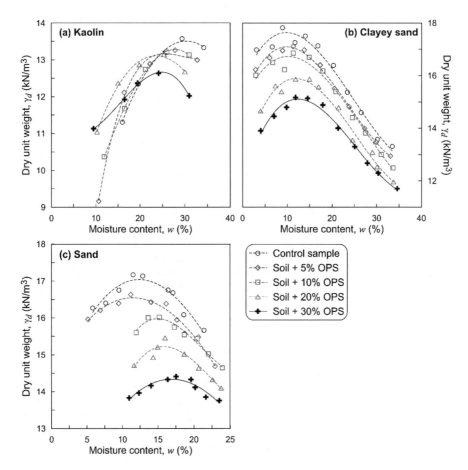

Fig. 2. a-c Compaction curves for the soil-OPS composites prepared using, (a) kaolin, (b) clayey sand, and (c) sand.

As expected the dry unit weights (γ_d) values increased with increase w up to reach the optimum point beyond which further increase in w resulted in decreasing γ_d values for each sample tested. Overall comparison of the w-γ_d results showed typical

Table 3. Outcomes of the compaction test for Kaolin-OPS composites

Soil type	Compaction parameters	OPS content (%)				
		0	5 (12.07)	10 (24.13)	20 (48.27)	30 (72.40)
Kaolin	w_{opt} (%)	30.30	26.83	28.18	23.71	23.59
	$\gamma_{d,max}$ (kN/m^3)	13.76	13.68	13.49	13.38	12.71

The values in the parenthesis shows the OPS content expressed by volume of soil

Table 4. Outcomes of the compaction test for sand-OPS composites

Soil type	Compaction parameters	OPS content (%)				
		0	5 (13.15)	10 (26.29)	20 (52.58)	30 (78.87)
Sand	w_{opt} (%)	12.54	11.46	14.96	16.31	16.95
	$\gamma_{d,max}$ (kN/m^3)	17.36	16.91	16.19	15.48	14.61

The values in the parenthesis shows the OPS content expressed by volume of soil

Table 5. Outcomes of the compaction test for clayey sand-OPS composites

Soil type	Compaction parameters	OPS content (%)				
		0	5 (13.00)	10 (26.00)	20 (51.99)	30 (77.99)
Clayey sand	w_{opt} (%)	10.38	10.44	10.83	12.75	13.55
	$\gamma_{d,max}$ (kN/m^3)	17.86	17.26	16.96	16.02	15.32

The values in the parenthesis shows the OPS content expressed by volume of soil

decreasing trends of γ_d with increasing OPS content. That is, the clean soil (control sample) for a particular soil type will have the maximum γ_d values corresponding to w_{opt} and the sample containing the higher OPS content (30% in this study) shows the minimum γ_d value at w_{opt}. These trends or behavior are also in good comparison with the previous literature (Gungat et al. 2013; Jamellodin et al. 2010; Prabakar and Sridhar 2002). This clearly highlights the influence of OPS content on the decreasing γ_d values, which relates to the lower specific density or specific gravity of the OPS samples in comparison to that of soils.

On the other hand, the w_{opt} values are observed to decrease with increase in OPS content. This phenomenon is well associated with the water absorption capability of the surface fiber in the OPS samples. The increase OPS content accelerates the water absorption of the soil-OPS composites, particularly on the surface fibers, thereby reducing the moisture content indirectly (Jamellodin et al. 2010; Prabakar and Sridhar 2002).

No significant variation in w_{opt} was observed in case of kaolin-OPS composites, with w_{opt} and $\gamma_{d,max}$ values utmost varying 23–30% and 12.5–13.5 kN/m^3 respectively. This minimal variation in the w_{opt} and $\gamma_{d,max}$ values perhaps relates to the use of kaolin – clay sized particles, used in these composites. The finer clay particles are expected to blend relatively well with the porous surface (or texture) of the OPS samples during hand mixing. However, in case of sand and clayey sand samples, the relatively larger particle size of sand may have been trapped or filtered by the fine porous texture of the OPS, thereby preventing them from filling the pores. These pores, when the samples are not fully saturated – which is certain in case of compaction tests, is filled with air and/or water.

The presence of air will certainly decrease the $\gamma_{d,max}$ measurements, while the pores accommodating more moisture will result in an increased w_{opt} measurements. This statement is very clearly demonstrated by the results for the soil-OPS composite prepared using sand, where the increasing OPS contents result in increasing w_{opt} – attributed to more moisture occupying the porous surface texture of OPS, while decreasing $\gamma_{d,max}$ values – attributed to air occupying the remaining pores of OPS. While, clayey sand – with particles sizes of sand and clay, show a mixed variation of kaolin and sand discussed above. That is, the increasing OPS contents result in increasing w_{opt}, while decreasing $\gamma_{d,max}$ values, but the rate of variation in the w_{opt} and $\gamma_{d,max}$ values are lesser when compared to that of sands.

4.2 Effect of OPS Content on CBR Test Result

Following the above discussions, it was rather crucial in the next step to evaluate the effect of the porous surface texture of OPS when the soil-OPS composite is subjected to saturation, or soaking in other words. This was achieved using the CBR set up and procedure as discussed in the methodology section of this paper. The samples were soaked in water for a period of 96 h prior to subjecting them to CBR testing. The swelling readings of the specimen at this stage was taken at a duration of every 24 h. Figure 3a–c shows the total swelling measured during the soaking of all three soil-OPS composites prepared in this study. The observation overall suggests that the inclusion of OPS in soil results in an increase in the swelling of the soil-OPS composite with time. Apparently, the increasing swelling is also relating to the water absorption capacity and/or behavior of OPS, or organic matter in general (Ghavami et al. 1999).

Subjected to soaking, the surface fibers on the OPS is likely to absorb water, filling its porous texture and thus exhibiting a volumetric expansion. This also supports the state-of-the-art practice of water repellent treatment applied on the natural fiber in filed prior to the soil reinforcement (Ghavami et al. 1999; Rahman et al. 2007). From the experimental results obtained in this study, the maximum swelling was observed in the initial 24 h of soaking, which is evident in the soil-OPS composites as well. The maximum swelling was measured up to 15 mm for kaolin composite, 4 mm for sand composites, and 6 mm for composites prepared using clayey sands. The higher swelling in kaolin obviously attributes to the swelling associated with the kaolinite mineral, while the lower swelling in sand and clayey sands is associated with low swelling potential of the soil samples.

Figure 4 shows the load bearing capacity (CBR) of the soil-OPS composites prepared using unsoaked and soaked conditions. The unsoaked samples showed trend closely resembling a linear variation in the CBR values of the composite with increasing OPS contents. The CBR values as high as 45% was measured for sand-OPS composite with 30% OPS content which is relatively good compared the control sample (clean sand) that measured only 24.25% - signifying almost 2 folds improvement in the CBR values in both unsoaked and soaked conditions. Composites prepared

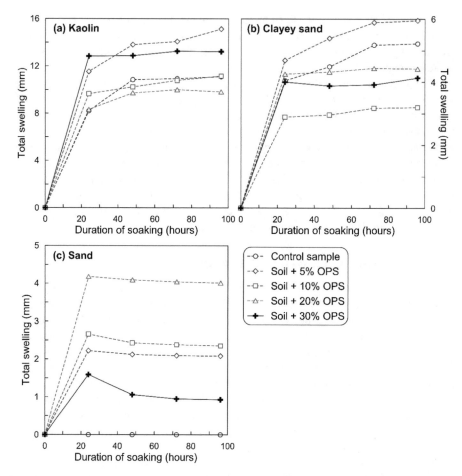

Fig. 3. a-c Total swelling measured during the soaking of the soil-OPS composites prepared in the CBR mold using, (a) kaolin, (b) clayey sand, and (c) sand.

using kaolin and clayey sand also showed improvement in the CBR values though not as significant as in the sands. The improvement in soil-OPS composite can be well explained using the concept of soil reinforcement – that the soil which is apparently weak in tension is improved by providing tensile reinforcement (Chauhan et al. 2008). In similar lines, the tensile resistance of the OPS as evinced in the literatures focusing OPS-concrete, is perhaps one of the key contributors to improvement observed in the CBR values (Okafor 1988; Mannan and Ganapathy 2001; Mannan and Ganapathy 2002; Ndoke 2006; Olanipekun et al. 2006; Teo et al. 2007).

However more in-depth research is of immediate requirement to confirm this statement, and also to evaluate the mechanism governing this improvement in the CBR or resistant in general. One known mechanism for example explains the contributions of the irregularities and uneven surface texture of the OPS to the improvement of strength and resistance of OPS towards applied loads (Gungat et al. 2013). In relation

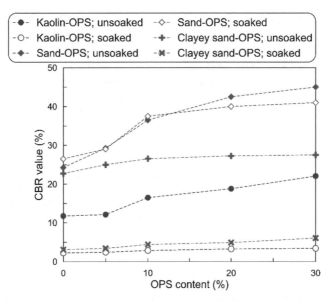

Fig. 4. Load bearing capacity soil-OPS composite under soaked and unsoaked conditions

to soaking effect, it is iterated in literature (Ghavami et al. 1999) that the organic fibers are sensitive by the moisture variations, likely to affect the interaction of reinforcing fiber and soil. Though the results obtained in this study compares well the results presented by Lawton and Fox (1992) on fiber reinforced soils, a more specific research concentrating on OPS reinforcement of soils is rather crucial.

5   Conclusions

This paper presents a pilot laboratory investigation to assess the influence of OPS content on the compaction and CBR behaviors of the soil-OPS composites. Overall comparison of the compaction (w-γ_d) results showed typical decreasing trends of γ_d with increasing OPS content signifying the influence of OPS content on the decreasing γ_d values, which relates to the lower specific density or specific gravity of the OPS samples in comparison to that of soils. While the water absorption capability of the surface fiber in the OPS samples is most likely to control the w_{opt} values. The results suggest the porous surface texture of the OPS as one of the key parameter controlling the $\gamma_{d,max}$ and w_{opt} measurements. The pores on the OPS surface, when the samples are not fully saturated – which is certain in case of compaction tests, is filled with air and/or water. Thus, presence of air is believed to decrease the $\gamma_{d,max}$ measurements, while the pores accommodating more moisture result in an increased w_{opt} measurements. Nonetheless the soil particle size mostly influences the rate of variation in the w_{opt} and $\gamma_{d,max}$ values of soil-OPS composite.

The swelling measured in the soil-OPS samples well relates to the inclusion of OPS in soil, with utmost emphasis on the water absorption capacity and/or behavior of OPS, or organic matter in general. The swelling associated with the volumetric expansion in OPS further depends on the type of soil and its swelling behaviors, as particularly observed and compared between the kaolin and sandy soils used in this study. The improvements in CBR values is associated to the tensile reinforcements provided by the OPS samples, and the rate of improvement is expected to relate with contributions offered by the tensile strength of individual OPS samples. In concluding this paper, an in-depth research is suggested to map and confirm contributions of individual OPS samples and to evaluate mechanisms governing improvements in the CBR values.

Acknowledgments. The first author gratefully acknowledge the financial support in terms of scholarship for his doctoral program from the Ministry of Higher Education Malaysia through the Fundamental Research Grant Scheme (FRGS/1/2015/TK08/MUSM/03/1). The support of Sime Darby Plantation Berhad, Malaysia is also gratefully acknowledged for support in terms of providing access the Oil Palm Shell samples that are used in this study.

References

Abdullah, N., Sulaiman, F.: The oil palm wastes in Malaysia. In: Biomass Now-Sustainable Growth and Use. InTech (2013). https://doi.org/10.5772/55302

Adeboje, A., et al.: Stabilization of lateritic soil with pulverized palm kernel shell (PPKS) for road construction. Afr. J. Sci. Technol. Innov. Dev. (2017). https://doi.org/10.1080/20421338.2016.1262100

Alengaram, U.J., et al.: Effect of aggregate size and proportion on strength properties of palm kernel shell concrete. Int. J. Phys. Sci. **5**(12), 1848–1856 (2010)

Alengaram, U.J., et al.: Utilization of oil palm kernel shell as lightweight aggregate in concrete – a review. Constr. Build. Mater. (2013). https://doi.org/10.1016/j.conbuildmat.2012.08.026

Awalludin, M.F., et al.: An overview of the oil palm industry in Malaysia and its waste utilization through thermochemical conversion, specifically via liquefaction. Renew. Sustain. Energy Rev. (2015). https://doi.org/10.1016/j.rser.2015.05.085

Chauhan, M.S., et al.: Performance evaluation of silty sand subgrade reinforced with fly ash and fibre. Geotext. Geomembr. (2008). https://doi.org/10.1016/j.geotexmem.2008.02.001

Economic Planning Unit: Eleventh Malaysia Plan, 2016–2020 (2015). Accessed http://www.epu.gov.my/en/rmk/eleventh-malaysia-plan-2016-2020

Ekeocha, N., Agwuncha, F.: Evaluation of palm kernel shells for use as stabilizing agents of lateritic soils. Asian Trans. Basic Appl. Sci. **4**(2), 1–7 (2014)

Ghavami, K., et al.: Behaviour of composite soil reinforced with natural fibres. Cement Concr. Compos. (1999). https://doi.org/10.1016/s0958-9465(98)00033-x

Gray, D.H., Al-Refeai, T.: Behavior of fabric-versus fiber-reinforced sand. J. Geotech. Eng. (1986). https://doi.org/10.1061/(asce)0733-9410(1986)112:8(804)

Gungat, L., et al.: Effects of oil palm shell and curing time to the load-bearing capacity of clay subgrade. Procedia Eng. (2013). https://doi.org/10.1016/j.proeng.2013.03.063

Ibrahim, A.: News article: New use for natural rubber, on STAR online (2018). https://www.thestar.com.my/opinion/letters/2018/03/21/new-use-for-natural-rubber/. Accessed 27 Apr 2018

Jamellodin, Z. et.al.: The effect of oil palm fibre on strength behavior of soil. In: Proceedings of the 3rd International Conference on Southeast Asian Natural Resources and Environmental Management (SANREM) 2010, Kota Kinabalu, Sabah, August 2010

Jeffay, J.: News article: Scottish council plans to fill potholes with recycled plastic waste, on The Scotsman (2018). https://www.scotsman.com/news/environment/scottish-council-plans-to-fill-potholes-with-recycled-plastic-waste-1-4702142. Accessed 27 Apr 2018

Khankhaje, E., et.al.: Comparing the effects of oil palm kernel shell and cockle shell on properties of pervious concrete pavement. Int. J. Pavement Res. Technol. (2017). https://doi.org/10.1016/j.ijprt.2017.05.003

Kumar, R., et al.: Engineering behavior of fibre-reinforced pond ash and silty sand. Geosynth. Int (1999). https://doi.org/10.1680/gein.6.0162

Lawton, E.C., Fox, N.S.: Field experiments on soils reinforced with multioriented geosynthetic inclusions (1992)

Li, J., et al.: Effect of discrete fibre reinforcement on soil tensile strength. J. Rock Mech. Geotech. Eng. (2014). https://doi.org/10.1016/j.jrmge.2014.01.003

Loh, S.K.: The potential of the Malaysian oil palm biomass as a renewable energy source. Energy Convers. Manag. (2017). https://doi.org/10.1016/j.enconman.2016.08.081

Mannan, M.A., Ganapathy, C.: Mix design for oil palm shell concrete. Cem. Concr. Res. (2001). https://doi.org/10.1016/S0008-8846(01)00585-3

Mannan, M.A., Ganapathy, C.: Engineering properties of concrete with oil palm shell as coarse aggregate. Constr. Build. Mater. (2002). https://doi.org/10.1016/S0950-0618(01)00030-7

Mohammed, M.A.A., et al.: Hydrogen rich gas from oil palm biomass as a potential source of renewable energy in Malaysia. Renew. Sustain. Energy Rev. (2011). https://doi.org/10.1016/j.rser.2010.10.003

MPOB (2016), Monthly production of oil palm products summary for the month of December 2016. http://bepi.mpob.gov.my/index.php/en/statistics/production/168-production-2016/747-production-of-oil-palm-products-2016.html. Accessed 16 Oct 2018

Ndoke, P.: Performance of palm kernel shells as a partial replacement for coarse aggregate in asphalt concrete. Leonardo Electron. J. Pract. Technol. 5(9), 145–152 (2006)

Nomanbhay, S.M., Palanisamy, K.: Removal of heavy metal from industrial wastewater using chitosan coated oil palm shell charcoal. Electron. J. Biotechnol. (2005). https://doi.org/10.2225/vol8-issue1-fulltext-7

Okafor, F.O.: Palm kernel shell as a lightweight aggregate for concrete. Cem. Concr. Res. (1988). https://doi.org/10.1016/0008-8846(88)90026-9

Okoroigwe, E.C., et al.: Characterization of palm kernel shell for materials reinforcement and water treatment. J. Chem. Eng. Mater. Sci. 5(1), 1–6 (2014). https://doi.org/10.5897/JCEMS2014.0172

Okpala, D.C.: Palm kernel shell as a lightweight aggregate in concrete. Build. Environ. (1990). https://doi.org/10.1016/0360-1323(90)90002-9

Olanipekun, E.A., et al.: A comparative study of concrete properties using coconut shell and palm kernel shell as coarse aggregates. Build. Environ. (2006). https://doi.org/10.1016/j.buildenv.2005.01.029

Prabakar, J., Sridhar, R.S.: Effect of random inclusion of sisal fibre on strength behavior of soil. Constr. Build. Mater. (2002). https://doi.org/10.1016/s0950-0618(02)00008-9

Rahman, M.M., et al.: Influences of various surface pretreatments on the mechanical and degradable properties of photografted oil palm fibers. J. Appl. Polym. Sci. (2007). https://doi.org/10.1002/app.26481

Rokade, S.: Use of waste plastic and waste rubber tyres in flexible highway pavements. In: Proceedings of the International Conference on Future Environment and Energy, IPCBEE 2012, Singapore, vol. 28, pp. 105–108 (2012)

Shafigh, P., et.al.: Oil palm shell as a lightweight aggregate for production high strength lightweight concrete. Constr. Build. Mater. (2011a). https://doi.org/10.1016/j.conbuildmat.2010.11.075

Shafigh, P. et.al.: A new method of producing high strength oil palm shell lightweight concrete. Mater. Des. (2011b). https://doi.org/10.1016/j.matdes.2011.06.015

Sharma, V. et.al.: Enhancing compressive strength of soil using natural fibers. Construction and Building Materials (2015). https://doi.org/10.1016/j.conbuildmat.2015.05.065

Shuit, S.H., et al.: Oil palm biomass as a sustainable energy source: a Malaysian case study. Energy (2009). https://doi.org/10.1016/j.energy.2009.05.008

Sriraam, A.S.: Effect of palm oil on the compressibility and hydraulic properties of kaolin. Ph.D. thesis, submitted for external examination, under review. Monash University Malaysia (2018)

Sumathi, S., et al.: Utilization of oil palm as a source of renewable energy in Malaysia. Renew. Sustain. Energy Rev. (2008). https://doi.org/10.1016/j.rser.2007.06.006

Teo, D.C.L., et al.: Lightweight concrete made from oil palm shell (OPS): structural bond and durability properties. Build. Environ. (2007). https://doi.org/10.1016/j.buildenv.2006.06.013

The Sun Daily: News article: Malaysia's new rubberised road technique a world-first, on STAR online (2017). http://www.thesundaily.my/news/2017/06/05/malaysias-new-rubberised-road-technique-world-first. Accessed 27 Apr 2018

Yean, G.P., ZhiDong, L.: A study on Malaysia's palm oil position in the world market to 2035. Renew. Sustain. Energy Rev. (2012). https://doi.org/10.1016/j.rser.2014.07.059

Yetimoglu, T., Salbas, O.: A study on shear strength of sands reinforced with randomly distributed discrete fibers. Geotext. Geomembr. (2003). https://doi.org/10.1016/s0266-1144(03)00003-7

Yew, M.K., et al.: Effects of oil palm shell coarse aggregate species on high strength lightweight concrete. Sci. World J. (2014). https://doi.org/10.1155/2014/387647

Multi-modal Transportation System Using Multi-functional Road Interchanges

Aušrius Juozapavičius[1]([⊠]) and Stanislovas Buteliauskas[2]

[1] Engineering Management Department, The General Jonas Žemaitis Military
Academy of Lithuania, Vilnius, Lithuania
ausrius.juozapavicius@lka.lt
[2] Pinavia Consortium, Vilnius, Lithuania
stanislovas.buteliauskas@pinavia.com

Abstract. Growing world population and increasing urbanization continues to stress cities and negatively impact people mobility. Morning and evening peaks of traffic congestion threaten to merge into a continuous never-ending congestion. Two main causes create this problem: an over-concentration of working places in city centers and a dis-proportionately large share of private cars in the traffic. Ownership and use of private vehicles are naturally encouraged by a typical layout of modern cities with living areas in suburbs which are dissolving passenger flows and making public transport ineffective there. A classical Park & Ride system is designed to attract and eliminate private cars coming from suburbs and transfer passengers to public transport for further travel towards the city center. However, most of these systems have limitations - if the parking space is large enough it requires a lot of real estate and it is difficult to access and exit, and if it is small enough to be convenient it does not have a sufficient effect. We present several innovative multifunctional road junctions which can act both as intersections of continuous flow and also utilize their occupied plot of land at the same time therefore becoming natural Park & Ride systems. All presented junctions have only two levels but traffic flows do not intersect. Their capacity is not limited by design. They enable non-interrupted flows in all directions as well as a possibility to make a U-turn. Most importantly, the junctions enable an easy access to a free area inside them. Thus the area can be utilized either for parking spaces, logistics or business centers, and public transport hubs with easy transfer capabilities between different modes of transport. Depending on the design one such junction can easily accommodate 20 thousand vehicles and create as many new working places. Placing the junctions on major intersections would both decentralize a city and create an effective infrastructure for multi-modal travel. Additionally, ITS can be easily employed to manage traffic flows and concentrate passengers creating a smart solution for city mobility. The multi-modal transportation system does not require any major changes of the road infrastructure inside a city and it is therefore a good choice both for a new or an old growing city.

© Springer Nature Switzerland AG 2019
S. El-Badawy and J. Valentin (Eds.): GeoMEast 2018, SUCI, pp. 133–142, 2019.
https://doi.org/10.1007/978-3-030-01911-2_12

1 Introduction

Traffic congestion is one of the major problems in most cities and it costs an average driver as much as 10,000 USD per year according to the INRIX 2017 Global Traffic Scorecard. Even though these numbers may be over-estimated and based on a wrong base line for comparing travel speeds (Cortright 2010), a push towards a more effective transportation system remains a pressing issue for any city (Small and Verhoef 2007). World's urban population will increase by estimated 2.5 billion during the next three decades (World Urbanization Prospects 2014) and the burden on the existing transportation systems will increase as well. Overpopulation of cities leads to decentralization efforts (Abougabal 2016) and the horizontal growth of cities inevitably increases the quantity of private cars as the ever expanding territory cannot be effectively served by the public transport. Therefore traffic congestion becomes an integral part of life in any city and cannot be eliminated completely (Downs 2005). Despite this fact it is still possible and desirable to reduce the congestion using a combination of various means such as congestion charging, car pooling and car sharing, expansion of road networks, building bus rapid transit (BRT) routes (e.g. (Olawepo 2013)), employing intelligent transportation systems, or eventually using driverless cars. Each of these and other elements has its limitations and neither of them is sufficient. This article presents an additional tool to the toolbox of a city transport engineer and policy maker – an innovative road junction which could serve as an integrating element between the private car and a public transport.

2 Multi-functional Junctions

When total daily traffic in a junction exceeds 35–40 thousand vehicles, a graded junction might be considered ("Traffic Management Guidelines" 2003) from Diamond interchange being one of the simplest to the 4-level stacked interchange being most costly. All of them have just one purpose – to eliminate crossings of flows from different roads on the same level, and therefore we call them single-functional junctions. Grade separation together with high speed left-turn lanes implies that these junctions require more land than simple at-grade intersections. Usually this land cannot be used for anything else than roads, but some junction layouts create sufficiently large empty spaces within them, and they could in principle be used for some additional infrastructure. One example could be the Cloverleaf junction with a fuel station or a parking lot in one of its leaves. Even though such structures have been observed (e.g., in Moscow) they are not very popular, because they are quite small and rather difficult (and not very safe) to reach. There are however designs where this additional functionality of a junction becomes economically feasible. One of them is Pinavia.

2.1 4-Way Pinavia Junction

The 4-way Pinavia junction (EP 1 778 918 B1 2012) was invented by trying to optimize traffic flows in a roundabout and by adding 4 small tunnels for each entering road to first cross the circular lanes and then merge them from the inside (Fig. 1).

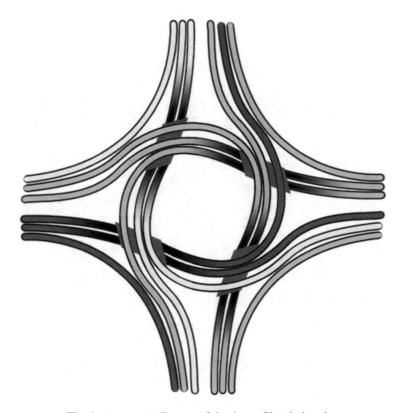

Fig. 1. A concept diagram of the 4-way Pinavia junction

Each tunnel can be independently replaced by an overpass to make a better use of the terrain. Depending on the projected capacity each roadway can be designed with as many lanes as needed. The tunnels (or overpasses) are built for only half of each road and they are also relatively short making their total cost of the same order of magnitude as for the usual Cloverleaf junction. The design makes it possible to have all road curves of the same radius and therefore keep a constant driving speed throughout the junction. Consequently, the size of the junction grows with the projected driving speed and reaches 600–700 m in diameter for 70 km/h speeds. It becomes unsuitably large above these speeds. On the other end it cannot be smaller than 300 m in diameter because the slope of the lane going from one tunnel and climbing above the next one is a limiting factor. However, this limit can be overcome by creating an elliptical version of the Pinavia with the longer axis along the major road. In that case the size of the junction in the direction of the minor road can be decreased down to 150 m for 50 km/h driving speed on that road and even more for smaller speeds.

The size of the symmetrical Pinavia is clearly a drawback until one considers the possibility to use its center. An extra lane on the left of an entry road makes it possible to enter and leave the center area without additional tunnels or overpasses (Fig. 2). This continuous circular area of 500–600 m in diameter and easily accessible from all four

Fig. 2. A possible infrastructure layout inside the Pinavia junction

roads may be used for various-purpose buildings, for large-volume parking lots, and especially – for a diverse public transport infrastructure. The layout is optimal for BRT routes. Also, it is possible to add a small-radius roundabout in the center giving the junction a U-turn functionality.

In order to better emphasize advantages of this junction we present several popular graded junctions and their parameters in Table 1. The junctions are included in the order of increasing building costs. We also included in the comparison a recently invented Double Crossover Merging Interchange (US 20130011190 A1 2013) as it might become a good alternative to the ubiquitous Diamond. The building costs are usually an indicator of the needed plot of land, and Pinavia would not stand well in that

Table 1. Comparison of selected 4-road graded interchanges

Type	Capacity, veh/h		Conflict points	U-turn	Multi-functionality
	Total	Left turn			
Diamond	5400	1500	12	Yes	No
Cloverleaf	17400	3000	4	Yes	Limited
Turbine	19200	4800	0	No	No
DCMI	19200	4800	0	Yes	No
Pinavia	19200	4800	0	Yes	Yes
Stacked 4-level	19200	4800	0	No	No

Notes: DCMI – Double Crossover Merging Interchange. Each road has 3 lanes: one for the left turn, one for the right turn, and one for going through.

light. However, the ability to commercialize its easily accessible center area should also be included in its economical evaluation together with its possible effect on the city traffic congestion as it will be pointed out later.

Due to its considerable size the junction cannot be used inside a city center, and due to speed limits in cannot be used for highways. An optimal location for the junction is therefore in the intermediate zone around a city – where arterial roads tend to have speeds up to 70 km/h, and the area is less urbanized. Closer to the center of a city the circular shape of the junction may not fit well with the usual rectangular network of the city roads. In that case there is an interesting alternative – to square it out as in the next example.

2.2   Multi-modal Multi-functional 2-Level Junction for One-Way Roads

Many cities with an infrastructure of parallel streets tend to have irregular distances between the crossings. This makes it impossible to implement a green wave of traffic lights in both directions simultaneously. Indeed, for a given driving speed v (m/s) and a traffic lights cycle of t seconds, a green wave in both directions will be possible if the crossings are separated by $L = v \times t/2$ meters. In that case when the platoon of cars from the first crossing reaches the second one just when its traffic lights become green, the platoon moving in the opposite direction reaches the first crossing at exactly the moment when its traffic lights also become green. Otherwise only part of the returning traffic will be able to pass the crossing (Fig. 3) during the green cycle. Given $v = 50$ km/h, and traffic lights cycle lengths between 60 and 80 s, we obtain optimal

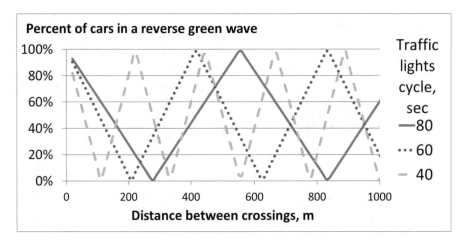

Fig. 3. A share of cars participating in a green-wave depending on the distance between crossings in a two-way street. Selected driving speed $v = 50$ km/h.

distances between the crossings to be in the range between 400 and 550 m. However, many cities tend to have blocks or at least pedestrian crossings separated by 200–300 m. Unless the cycles of traffic lights are set to a very small period (40 s or less) it becomes impossible to have a green wave in both directions on a two-way street. In that case it is possible to convert streets into one-way streets with traffic direction alternating in the style of Manhattan or Barcelona. Even then, random distances between crossings make it complicated to create green waves on all the streets. However, a grade separation using simple bridges makes it possible to create an interesting structure which is both a grade-separated junction and a public transport hub at the same time (Fig. 4). We call it a Multi-modal multifunctional 2-level junction for one-way roads, because it has two functions – it is both an interchange and a parking

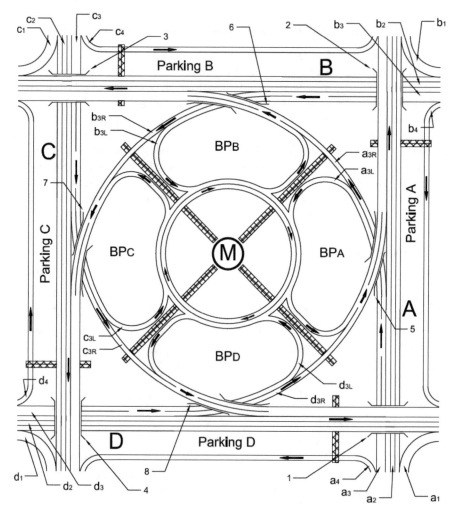

Fig. 4. Multi-modal multi-functional 2-level junction for intersections of one-way streets. The distance between the parallel streets (A to C) is 200–300 m. BP denotes bus parking, and hatched rectangles denote passage ways for pedestrians.

place, and it also allows easily changing the mode of transport. At the same time, we do not need traffic lights anymore.

Apart from buses this junction can have other conveniently accessible modes of transport integrated in it. A cross-section of a possible underground public transport hub inside the multi-functional junction is presented in Fig. 5. Here number 1 denotes a tram, 2 – tram rails, 3, 4, 7 – stairways for passengers, 5 – a metro car, 6 – its rail, 8 – an electric light train, 9 – its rail, 10 – an office building, 11 – a logistics terminal, 1PL and 2 PL – first level and second level car parking, a_2, d_2, b_2, c_2, a_5, c_5 – roadways, a_4, c_4 – a dedicated lane for public transport (e.g. BRT), b_3, d_3 – pedestrian lanes.

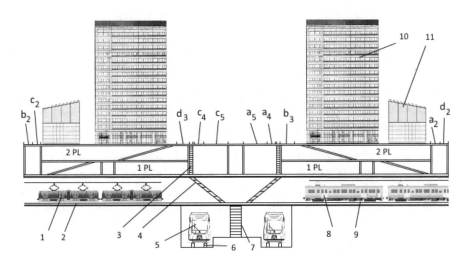

Fig. 5. A cross-section of a multi-functional junction

3   City Layout for Multi-modal Transportation System

The use of multi-functional junctions has implications for the whole city transportation system layout. First of all, these junctions are natural and very convenient Park & Ride places, and should be placed therefore in the intermediate zone of a city – before the traffic congestion starts. Even though many cities are establishing parking lots close to light-rail or metro stations, they usually have small effect because of their size: it is costly to make a large parking lot where other structures make more economic sense and they require additional road infrastructure to reach them. Usually such parking lots have places for 100–200 cars. Making people switch private cars for public transport already implies some resistance, because the convenience of the private space is lost. Making it complicated for them to perform the transfer creates an additional challenge. Indeed, a Danish study estimating the value of travel time indicates that travelers attribute 1.5 times more value to the time needed to switch travel modes than to the in-vehicle time and the act of mode change is additionally equivalent to a 6 min delay

(Fosgerau et al. 2007). Therefore, a parking lot inside a multi-functional junction such as Pinavia is superior to any by-the-road parking space as it is straightforward to reach, it is sufficiently large to accommodate 5–6 thousand cars on one level (and it may have 3 easily accessible levels) and there is a very short distance from any parked car to the public transport station or a terminal. Most importantly, the total area does not take up significantly more space than another junction of similar throughput capacity.

Apart from the Park & Ride infrastructure the multi-functional junctions can act as centers for a decentralized city by creating many new working places as they can be built on top of the parking lots. Easy access would make them economically attractive for real estate developers and a shift of working places away from the city center would alleviate the overall congestion even more.

A possible city layout incorporating multifunctional junctions is presented in Fig. 6. Number 1 denotes a ring road around the old town O, number 2 – a ring road of two one-way roads around the center part of the city C, and number 3 – a similar ring around the zone of new developments N. Number 4 denotes radial arterial streets. They are also consisting of two one-way roads. All intersections are suitable for multi-functional junctions denoted by number 5. Normally the old town zone would have limitations on private transport and no transit through it. Usually a growing city N puts more stress on its C zone, but the doubled one-way streets could create convenient easily accessible locations for offices, supermarkets, and businesses and even encourage some decentralization as depicted in Fig. 7, where businesses are relocated between the one-way streets. The letter J denotes locations for multi-functional junctions, and number 10 denotes the area for new residential buildings. The multi-functional junctions should be 3–5 km apart. They should have parking places for private cars, and hubs for public transport. Having parking places inside the junctions

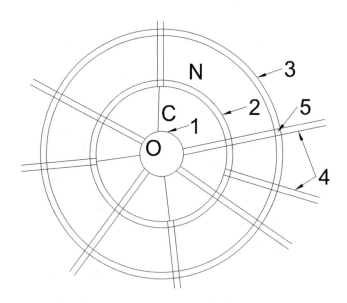

Fig. 6. An example of a city infrastructure layout

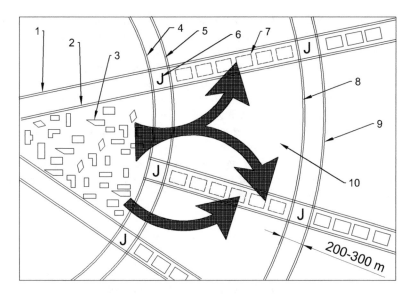

Fig. 7. A possible decentralization of a city. Black arrows denote new possible places for businesses

helps to eliminate the bottleneck problem inherent in other junctions. Namely, when a high capacity junction such as a 4-level stacked interchange or a Turbine is built, it almost always has larger capacity than the exiting road: e.g., it may take a major road of 3 lanes and merge it with a minor road of 2 lanes, and the resulting main road of 5 lanes gets then squeezed back into 3 lanes during the next several kilometers. If both roads were loaded to their full capacity then the resulting road would be overloaded and the drivers would have no alternative except to stand in line for their turn to move (Downs 2005). Meanwhile, having a parking lot inside a junction with a possibility to switch to a public transport and an ITS informing drivers of the possible alternative would eliminate some traffic from the road.

As a rule new fast-traffic infrastructures attract or even generate more traffic so that they become as congested as any other road (Litman 2001). Therefore, a single multi-functional junction will not make a difference. However, if they are distributed around the city they will not be generating any new traffic because of the symmetry, and the redirected traffic will be small and surpassed by the increased share of public transport users.

4 Conclusions

Most of graded junctions are placed so that they merge several lanes of two roads into a road with a smaller number of resulting lanes thus creating a bottleneck problem for traffic. This is not a drawback of the junctions themselves, but of the roads as they cannot have ever more lanes after each intersection. However, multi-functional junctions with integrated parking spaces make it possible to reduce the excess traffic – the traffic may be diverted into the parking using an ITS or some form of congestion charging, but mostly it will happen naturally if the public transport will be convenient and fast enough to attract the Park & Ride drivers and passengers.

The cost of multi-functional junctions presented in the article is comparable to the cost of other graded junctions but the benefits of the added functionality make the multi-functional junctions far superior.

The multi-functional junctions with easily accessible areas also create favorable locations for logistics centers or other businesses and may help to further decentralize cities and reduce their traffic congestion.

Using multi-functional interchanges on city ring-roads would create favorable conditions for multi-modal city transportation system and potentially reduce traffic congestion.

References

Abougabal, H.: Cairo requires long-term planning. MEED Bus. Rev. **60**(1), 28–29 (2016)

Buteliauskas, S.: EP 1 778 918 B1. European Patent Office, EU (2012)

Cortright, J.: Measuring Urban Transportation Performance a Critique of Mobility Measures and a Synthesis. Chicago (2010)

Downs, A.: Still Stuck in Traffic, Coping with Peak-Hour traffic Congestion. Brookings Institution Press (2005). https://www.brookings.edu/book/still-stuck-in-traffic/

Fosgerau, M., Hjorth, K., Lyk-Jensen, S.: The Danish value of time study: final report (2007)

Gingrich, M.A.S.: US 20130011190 A1. US (2013). http://www.google.com/patents/US20130011190

Litman, T.: Generated traffic: implications for transport planning. ITE J. (Inst. Transp. Eng.) **71** (2001)

Olawepo, R.A.: Facilitating urban mass transit through expanded transport planning and management: an example from Lagos, Nigeria. In: Proceedings of the Multidisciplinary Academic Conference, pp. 1–9 (2013)

Small, K.A., Verhoef, E.T.: The Economics of Urban Transportation. Routledge, New York (2007)

Traffic Management Guidelines: Department of Transport of Ireland, Dublin (2003). https://www.nationaltransport.ie/downloads/archive/traffic_management_guidelines_2003.pdf

World Urbanization Prospects (2014). https://esa.un.org/unpd/wup/publications/files/wup2014-highlights.pdf

Deformation Behavior for Different Layered Cushion Material After Freeze-Thaw Cycles

Guo-Quan Ding[1,2(✉)], Jun-Ping Yuan[1,2], and Qi Wang[1,2]

[1] Geotechnical Research Institute, Hohai University, Nanjing, China
dingguoquan2004@126.com
[2] Key Laboratory of Ministry of Education for Geomechanics and Embankment
Engineering, Hohai University, Nanjing, China

Abstract. Layered cushion material is widely used in concrete-faced rockfill dams and embankment of highway. The deformation of the cushion material is dominant to the safety of the concrete face or pavement. If the total or differential deformation of cushion material is too large, cracks may develop in concrete face or pavement, and even more destruction may happen to the whole structure. In cold area, the excessive deformation of cushion material may happen because of repeated frost heave and thaw collapse during freeze-thaw cycles. For better understanding on this excessive deformation and optimum design for cushion material, series laboratory tests are carried out on four types of samples with different layered structure. A self-developed one-dimensional freeze-thaw device is used to monitor the deformation of specimen during freeze-thaw cycles. Test results show that it's not always frost heave or thaw collapse during freeze-thaw cycles. All specimen show collapse during frost and heave while thaw in the first freeze-thaw cycle. Comparisons are conducted on the four types of layered samples. Comparison results show that the sample with a "loose top and dense bottom" structure deforms relatively small out of four types of layered samples. It also shows that the deformation of frost heave and thaw collapse tends to a stable value along with the freeze-thaw cycles increasing.

Keywords: Layered cushion material · Freeze-thaw cycles · Frost heave
Thaw collapse

1 Introduction

Frozen soils are widely distributed in China. The area of permafrost soil and seasonal frozen soil takes up about 21.5% and 53.5% of China land area respectively. Because frozen soils have special adverse engineering properties, they should be studied clearly to ensure the safety, durability and economic rationality of the whole project construction (Ma and Wang 2012).

When designing and building concrete faced rockfill dam or embankment of highway, cushion material is always set under the concrete face or pavement to provide a flexible support for them. In seasonal frozen area, cushion material will freeze and heave in winter because of low temperature, and it will lead to an upward displacement for concrete face or pavement. But in summer, the frozen cushion material will thaw

© Springer Nature Switzerland AG 2019
S. El-Badawy and J. Valentin (Eds.): GeoMEast 2018, SUCI, pp. 143–151, 2019.
https://doi.org/10.1007/978-3-030-01911-2_13

and collapse because of warming temperature, while it will result in a downward displacement for concrete face or pavement. During freeze-thaw cycles, repeated frost heave and thaw collapse of cushion material may directly cause bump failure or tension fracture developed in concrete face or pavement, and then the safety of the whole project may be threatened. So, the physical and mechanical properties of cushion material after freeze-thaw cycles are very important. But at present, acquaintance about them are not very enough, thus it is very necessary to make a deep research on them.

Yuan et al. (2013, 2015a) carried out a series of laboratory tests to research the deformation and strength of cushion material after freeze-thaw cycles. Impact of pore size distribution, fine particle content and gradation on cushion material properties were studied particularly (Yuan et al. 2014, 2015b, 2016, 2017). They found out that, heave deformation decreases with freeze-thaw times, while increases with initial water content and relative density. Furthermore, frost heave deformation varies with size and quantity of the pore. There exists a critical pore size, when the pore is smaller than the critical size, the smaller the pore and the more the small pores, the greater the frost heave.

An et al. (2013) pointed out that frost heave of coarse grained soil was mainly affected by the particle size, water content and groundwater level. Initial water content of frost heave and the ability to stop capillary water from rising should be taken into consideration when choosing the right cushion material. They put forward a method for choosing cushion material according to capillary water potential in layered soil based on the analysis of frost heave resistant mechanism of cushion.

In dam or embankment engineering construction, cushion material is always filled in layers. In this paper, freeze-thaw cycle tests for cushion material with different layered structure are carried out, the influence of layered structure on frost heave and thaw collapse deformation is researched, and the deformation changing rule with freeze-thaw cycle times is also analyzed.

2 Material and Experimental Methods

2.1 Material

The material used in this study is a kind of rock-silt mixture, which taken from the cushion of a certain concrete-faced rockfill dam. The basic physical properties for the material are listed in Table 1, and the grain-size distribution curve is shown in Fig. 1.

Table 1. Basic physical properties of test material

Maximum particle size (mm)	Specific gravity	Maximum dry density (g/cm^3)	Minimum dry density (g/cm^3)
20	2.68	2.31	1.98

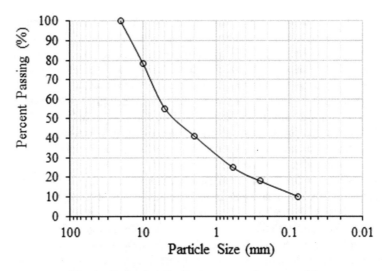

Fig. 1. Grain-size distribution curve of test material

2.2   Self-developed One-Dimensional Freeze-Thaw Device

To measure the deformation of cushion material specimen during freeze-thaw cycles, a self-developed one-dimensional freeze-thaw device is used, which is shown in Fig. 2. This device mainly consists of PVC circular tube, upper and lower plastic cover plate fixed by connecting bolts, dialgage, thermal resistor and digital reader etc. Two iron hoops are fitted around the PVC circular tube to avoid it exploded by frost-heaving sample.

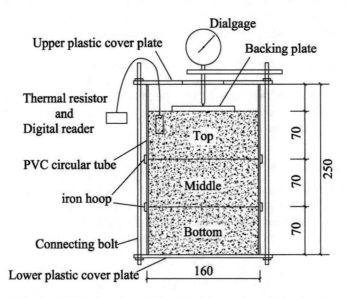

Fig. 2. Self-developed one-dimensional freeze-thaw device (mm)

PVC circular tube, the sample container, is 160 mm in external diameter, 250 mm in height and 3 mm in thickness. The cushion material is filled PVC tube in three layers. The height of each layer is 70 mm, so the total height of the sample is 210 mm. In tests, the relative density (D_r) of each layer is set to be different to simulate the layered structure of cushion material in engineering. Different layered structure means that the combination of D_r for three layers in one sample is different. For easy expression, three layers are named as top, middle and bottom individually.

During freeze-thaw cycles, the sample deformation is measured by the dialgage (0–30 mm), fastened to the stainless steel support. Temperature inside the sample is monitored by thermal resistor and digital reader.

Some auxiliary equipment are also used in this test, including refrigerator, which is the freezing installation, and foam box, which is the heat preservation device. In engineering practice, the thermal conduction in cushion material is one-dimension from top to bottom. To simulate this kind of thermal boundary in freeze-thaw cycle test, the sample is put into an open foam box, and the space between sample side and inner wall of the box is filled with foam particles, to avoid the heat transfer from the side and bottom of the sample as much as possible.

2.3 Freeze-Thaw Cycle Experiments

The purpose of the test in this paper is to research the impact of layered structure and freeze-thaw cycle times on specimen deformation of cushion material. So four kinds of samples with different layered structure, numbered as I, II, III and IV, are prepared to conduct freeze-thaw cycle experiments. As shown in Table 2, layered structure kinds include "loose top and dense bottom", "dense top and loose bottom", "dense top, bottom and loose middle" and "loose top, bottom and dense middle". The dense layer has a relatively large D_r, while the loose layer is opposite. D_r of three layers for each sample are listed in Table 2. Samples are compacted at water content of 6%. When samples are put into the refrigerator for freezing, the lowest temperature is controlled to be minus 10 °C.

For sample I, II and III, the freeze-thaw cycle times is 3, while for sample IV it is 22. Deformation of four samples during 3 times freeze-thaw cycles is recorded to comparatively analyze the influence of layered structure on specimen deformation. Besides, sample IV is chosen to research the development rule of freeze-thaw deformation with cycle times increasing from 1 to 22.

2.4 Sample Preparation

Samples are prepared according to the designed water content and layer D_r listed in Table 2. Natural water content of experimental cushion material is measured first. Then the mass of cushion material and filling water for one certain sample layer can be calculated. After that, they are mixed and compacted into the sample container, until the height of the layer reach the designed size (i.e. 70 mm).

Put the prepared sample into refrigerator for freezing. The real-time temperature of the sample can be measured by thermal resistor and digital reader. Specimen deformation is read by the dialgage from time to time.

Table 2. Freeze-thaw cycle test program

Sample number	Layered structure	Water content	Lowest freeze temperature	Freeze-thaw cycle times	Relative density D_r		
					Top	Middle	Bottom
I	loose top and dense bottom	6%	−10 °C	3	0.7	0.8	0.9
II	dense top and loose bottom	6%	−10 °C	3	0.9	0.8	0.7
III	dense top, bottom and loose middle	6%	−10 °C	3	0.9	0.7	0.9
IV	loose top, bottom and dense middle	6%	−10 °C	22	0.7	0.9	0.7

3 Results

3.1 Deformation of Cushion Material Samples During Freeze-Thaw Cycles

According to aforementioned method and program, freeze-thaw cycle tests for cushion material samples with different layered structure are carried out. The deformation of samples during freeze-thaw cycles are measured, and then plotted in Fig. 3.

Figure 3 shows that, four cushion material samples with different layered structure have a similar deformation law. In general, the sample will heave when freezing and collapse when thawing. The frost heave deformation of cushion material mainly comes from the volume dilatation of pore water in the process of freezing, and the thaw collapse deformation is mainly due to the volume shrinkage in the process of ice melting into water.

Also shown in Fig. 3, there exists a phenomenon that, for all four samples, they will not heave but collapse for the first time freezing, and not collapse but heave for the first time thawing. It means that it is not always frost heave or thaw collapse during freeze-thaw cycles. It is perhaps because that: (1) in the early stage of frost heaving, the freezing of pore water may lead to particle dislocation and rearrangement in cushion material, then a certain volume reduction may develop; (2) in the early stage of thaw collapsing, temperature increasing and ice melting mainly happen in the surface layer of the sample, but the frost heaving may still develop inside the sample, then the heaving deformation may be still ongoing.

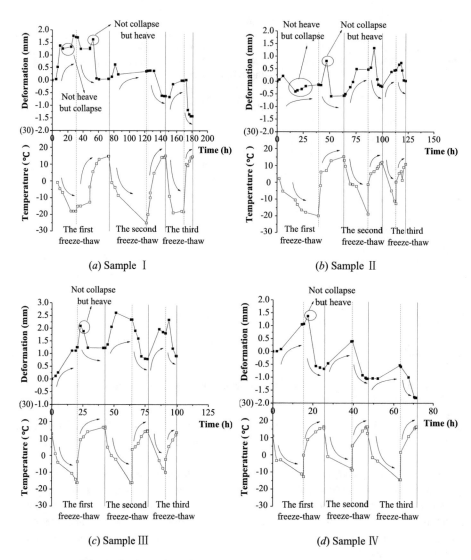

Fig. 3. Deformation of sample I ~ IV during 3 freeze-thaw cycles

3.2 Comparison on Frost Heave Deformation of Cushion Material Samples

In the nth freeze-thaw cycle, there are three characteristic values of the sample deformation, which are: $\delta_0{}^n$, the deformation at the beginning of this cycle; $\delta_h{}^n$, the deformation at the minimum temperature in this cycle; and $\delta_c{}^n$, the deformation at the maximum temperature in this cycle. The relative value $(\delta_h{}^n - \delta_0{}^n)$ and $(\delta_h{}^n - \delta_c{}^n)$ can be respectively regarded as the frost heave and thaw collapse of the sample for the nth freeze-thaw cycle.

To compare the frost heave property of four samples with different layered structure, the variation curves of frost heave deformation for all samples changing with
freeze-thaw cycle times (0–3) are plotted on Fig. 4. From this figure, the following
rules can be obtained:

(1) For the first freeze-thaw cycle, sample I has the largest frost heave, about 1.75 mm,
sample III and IV takes a middle one, and sample II's frost heave is minimal and

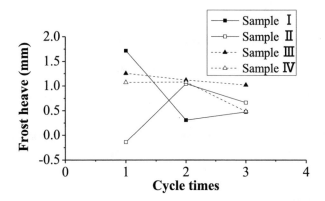

Fig. 4. Frost heave of cushion material samples with different layered structure

minus. The frost heave of sample II is minus, which do not mean this sample don't
heave, but it just caused by the phenomenon of "not heave but collapse". That is to
say, the collapse deformation is greater than the heave deformation.

(2) The frost heave of samples with different layered structure differs from one another.
Altogether, the frost heave of sample II increases with freeze-thaw cycle times. On
the contrary, the frost heave of sample I, III and IV decreases with freeze-thaw
cycle times. Also, the frost heave of sample I is overall less than the other two.

On the whole, sample I with a layered structure of "loose top and dense bottom"
may have a relatively small frost heave during the process of freeze-thaw cycles. So,
from the view of reducing frost heave, the better layered structure of cushion material
in engineering is "loose top and dense bottom".

3.3   Deformation Change Law During Multiple Freeze-Thaw Cycles

To research the change law of sample deformation during multiple freeze-thaw cycles,
sample IV with a layered structure of "loose top, bottom and dense middle" is chosen
to test with 22 times freeze-thaw cycles. The frost heave and thaw collapse deformation
is plotted in Fig. 5.

As shown in Fig. 5, frost heave and thaw collapse deformation doesn't vary
monotone but fluctuates along with freeze-thaw cycle times increasing. There exists a
mean vale, about 1.25 mm, around which the frost heave and thaw collapse deformation fluctuates. With the increasing of freeze-thaw cycle times, the amplitude of

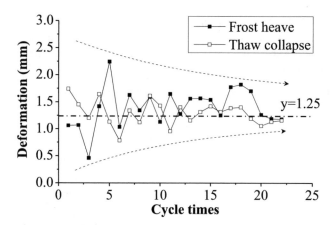

Fig. 5. Frost heave and thaw collapse deformation changing with freeze-thaw cycle times

fluctuation decreases, and the frost heave and thaw collapse deformation will tend to a relatively stable value, which is almost the mean value.

4   Conclusions

A self-developed device is designed and used to conduct freeze-thaw cycle tests on four samples with different layered structure. During freeze-thaw cycles the deformation of specimen is measured. The influence of layered structure on frost heave property and the change rule of sample deformation with freeze-thaw cycle times are analyzed. The main conclusions are as follows.

(1) It is not always frost heave or thaw collapse during freeze-thaw cycles. All specimen show collapse during frost and heave while thaw in the first freeze-thaw cycle.
(2) The sample with a "loose top and dense bottom" structure deforms relatively small out of four types of layered samples. The better layered structure of cushion material in engineering should be "loose top and dense bottom".
(3) The deformation of frost heave and thaw collapse tends to a stable value along with the freeze-thaw cycle times increasing.

Acknowledgments. This research was supported by the National Natural Science Fund of China (Grant No. 51378008), the National Key R&D Program of China (Grant No. 2017YFC0404804, 2017YFC0404801), and the Fundamental Research Funds for the Central Universities (Grant No. 2017B20614).

References

An, P., Xing, Y.C., Zhang, A.J.: Study of design method and numerical simulation for anti-frost heave cushion of canal. Rock Soil Mech. **34**(sup. 2), 257–264 (2013)

Ma, W., Wang, D.Y.: Studies on frozen soil mechanics in China in past 50 years and their prospect. Chin. J. Geotech. Eng. **34**(4), 625–640 (2012)

Yuan, J.P., Li, K.B., He, J.X., et al.: Heave deformation of cushion material for CFRD during freeze-thaw cycles. Sci. Technol. Eng. **13**(4), 1087–1090 (2013)

Yuan, J.P., Li, K.B., He, J.X., et al.: Discussion on frost heave deformation of cushion material based on pore distribution model. Rock Soil Mech. **35**(8), 2179–2183, 2190 (2014)

Yuan, J.P., Ding, P., Li, K.B., et al.: Test research on strength of cushion material during freeze-thaw cycles. Low Temp. Arch. Technol. (1), 19–22 (2015a)

Yuan, J.P., Wang, Q.L., Han, C.L.: The Influence of pore size distribution on cushion material strength after freezing-thawing cycle. Sci. Technol. Eng. **15**(21), 48–52 (2015b)

Yuan, J.P., Ding, P., Lin, Y.L.: Impact of fine particle content on frost-heave of cushion material. Water Resour. Hydropower Eng. **47**(2), 27–32 (2016)

Yuan, J.P., Lu, Y.P., Ji, S.R., et al.: Cushion material characteristics of frozen heaving and scale effect test. Sci. Technol. Eng. **17**(5), 284–288 (2017)

Evaluating Fouled Railway Ballast Using Ground-Penetrating-Radar

Chihping Kuo[✉]

Department and Institute of Civil Engineering and Environmental Informatics,
Minghsin University of Science and Technology, Xingfeng, Hsinchu, Taiwan
picnic.kuo@must.edu.tw

Abstract. Soil pumping phenomena comprise a wide issue in traditional railway structures. A newly built railway structure from the top down contains rails, slippers, ballasts and soil layer foundation. Currently about 85% of the railway structure is constructed traditionally in Taiwan as mentioned. With the infiltration of rainfall or groundwater, the soil foundation will become saturated. After repeated loading from passing, trains the soil will become even more saturated and squeezed into the voids inside the sub-ballasts or ballasts. The pumping effect is then initiated and the pumping paths start to scurry. The repeatedly force transfers through ballast to the saturated foundation, may create vacuum to draw phenomenon, called pump effect or mud pumping. It could lead to serious train derailment capsized. Present mud pumping detection method has to be performed during the non-operating time at night by visual. However, this approach may have omissions and shortcomings perspective concerns, and slow to find disasters during the rainy season. According to previous studied result, the pumped mud is characteristic of being quasi-liquefied, denser, and with exceed pore-water pressure. The loading from trains cause dynamic force to rise the pore-water pressure. Adopting the feature, a tubing-floater device is proposed to install on the interface between ballasts and soil layer foundation. No matter the ballast sinks or mud pumping up, the floater will be raised due to the pressure generated in the tube. The laboratorial experiment has been performed to prove the feasibility of the idea herein.

1 Introduction

Mud pumping is a common problem to rail bed and is often depicted as the penetrating of fine materials (usually from the subgrade, see Fig. 1) through the ballast. It may cause differential track settlements. The fine materials penetrated from the subgrade are usually visible, appearing as dried mud (during dry days) or as pool of muddy slurry during rainy spells (see Fig. 2). Tadatoshi (1997) proposed that mud pumping due to the force transferring through ballast to the saturated foundation, may create vacuum to draw phenomenon. A Japanese survey carried out in 1975 found that out of 17000 km of frequently used rails in Japan, more than 700 km suffer from mud pumping and regular track alignment and maintenance cycle lasted only 58 days on the average (Katsumasa 1978).

© Springer Nature Switzerland AG 2019
S. El-Badawy and J. Valentin (Eds.): GeoMEast 2018, SUCI, pp. 152–160, 2019.
https://doi.org/10.1007/978-3-030-01911-2_14

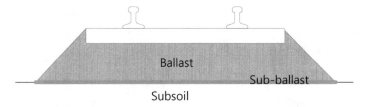

Fig. 1. Schematic diagram of railway foundation structure

Fig. 2. Pumping photo recorded on 2016.01.05 at Neiwan Branch, Hsinchu, Taiwan

2 The Mechanism of Pumping Phenomenon in Soil Foundation

Pumping phenomenon is generally considerate to be conducted by the migration of subsoil fine particles in the ballast layer (Ayres 1986; Aw 2004, 2007; Ghataora et al. 2006; Indraratna et al. 1997). Alobaidi and Hoare (1994, 1996) proposed that pumping of fine particles depends mainly on the pore water pressure developed at the interface between the subgrade and subbase/ballast layer. As the phenomenon occurring, the material would be carried out and then cavities under the ballast grows larger. It may induce serious train derailment cap-sized. Duong et al. (2014) adopted physical model for studying the migration of fine particles in the railway substructure. In the study, it was found that the development of pore water pressure in the subsoil is the key factor causing the migration of fine particles, hence, resulting in the creation of interlayer, as well as mud pumping. Tadatoshi (1997) proposed that pumping due to force transfers through ballast to the saturated foundation, may create vacuum to draw phenomenon (See Fig. 3).

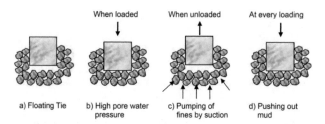

Fig. 3. Schematic diagram of pumping effect. (Takatoshi 1997)

3   Design Idea and Methodology

Since the mud-pumping behavior is associated with the migrating of fluid inside the void of soil particles and the fluid would saturate the mud, to detect the motion fluid and saturated soil particles will be helpful to realize the distribution of mud-pumping. Moreover, the uplift pressure generated by pumping mud due to the cyclic loading from train can be adopted for the mentioned detecting work. It is just to lead the pumping mud into a casing tube firstly, and then detect the height.

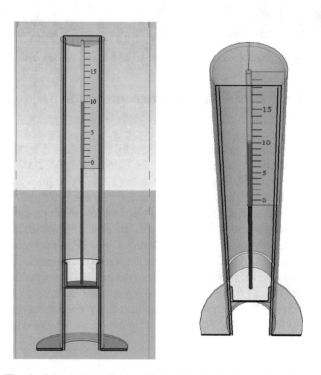

Fig. 4. Schematic of Tubing-Floater Device design methodology

The structural Schematic of Tubing-Floater Device design methodology is shown as Fig. 4. The entire set of the device includes casing tube and inside floater. The tube is adopted for leading the pumping mud to flow into the device and fill the chamber fully. The pressure generated from the mud inside the chamber will push the floater upward as the mud lift. For increasing the buoyancy, the floater is designed to be a raft-shaped. The outer surface of the floater and the inner surface of the tube should be close-fitting and slip smoothly. On the other hand, the tube should be a well cylinder. Otherwise, the floater will get stuck. A scale of distance that indicating the motion of floater is marked in the outside of the tube. The mentioned components are well-composed to be a Tubing-Floater Device. The complete product is shown in Fig. 5. It is made in high-strength Acrylic.

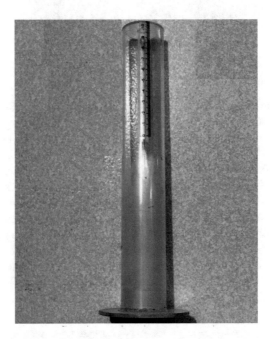

Fig. 5. Picture of Tubing-Floater Device

4   Practical Test of the Device

4.1   Experimental Model Setup

To simulate the field mud-pumping phenomenon, a small-scaled test was design and performed. A tank made in tempered glass with reinforcing metallic mesh was adapted for simulate the in-situ railway soil foundation. The lower layer was soil from field with mixed medium and fine particles (SM-ML with Effective grain size $D10 = 0.01$ mm). The upper layer was crashed gravel (GP with Effective grain size $D10 = 2$ cm), as shown in Fig. 6. Both materials were washed and then fully dried before test. The bottom of Tubing-Floater Device was placed at the interface of soil and gravel. A normal sprayer was adapted to simulate rainfall. It was deemed that entire foundation would be saturated while the water level reaches the surface of soil.

Fig. 6. The simulating setup for mud-pumping test

4.2   Simulating of Cyclic Loading

The cyclic loading from trains was simulated by using Proctor harmer. The Proctor compaction test (2007) is designed as a laboratory method of experimentally determining the optimal moisture content at which a given soil type will become most dense and achieve its maximum dry density. The original Proctor test, ASTM D698/AASHTO T99, uses a 4-inch-diameter (100 mm) mould which holds 1/30 cubic feet of soil, and calls for compaction of three separate lifts of soil using 25 blows by a 5.5 lb hammer falling 12 inches, for a compactive effort of 12,400 ft-lbf/ft^3. In this study, the sink versus the relationship of compaction loading pressure is performed. As Fig. 7 shows the different loading forces leads relative sinking. However, the quantity of pumping-mud those squeezed into the void of gravel cannot be observed directly. The curve shown in Fig. 8 represents the relationship between sinking and loading. It can be seen the fitting curve would be formed as parabola. Higher loading lead deeper sink, which means more mud-pumping potential will be triggered.

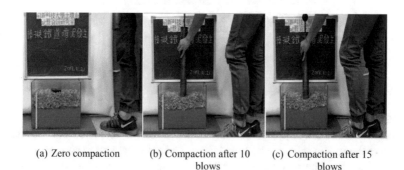

(a) Zero compaction   (b) Compaction after 10 blows   (c) Compaction after 15 blows

Fig. 7. The pictures recorded the sinking phenomenon due to the compacting

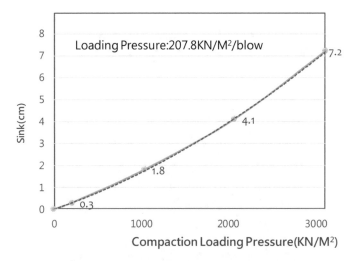

Fig. 8. The relationship between sinking and loading

4.3 Profile and Pattern of Pumped Mud

Some glue was mixed into the soil prior to the compaction. It was adapted for fixing the shape also the pattern generated by the pumped mud. The interesting point is the saturated mud could be successfully squeezed into the void of gravel as Fig. 9 shows, therefore the method can be well considerate to be a suitable method to simulate the mud-pumping behaviour.

Fig. 9. The simulating setup for mud-pumping test

4.4 Test Results

The Tubing-Floater Device was installed correctly. Two hammers made repeated loadings on gravel surface of the both side continuingly. After the cyclic loading, it can be found the device worked as expectation. The floater inside the tube uplift well even higher than the surface of gravel. It is because the surface level of the gravel already sank posterior to the compaction. No matter which quantity of loading was performed, even the mud was squeezed up to the surface, the level of floater would be still higher than the mud level, as shown in Fig. 10.

Fig. 10. Different view of the uplift floater by cyclic loading

4.5   Discussion

The relationship between the uplift height of floater and the sinking depth can be simplified to discussed by observing the sectional profile in Fig. 11. The distance of the boundaries in the tank is defined as parameter "D", the diameter of the chamber is defined as parameter "d", the height of the floater is defined as parameter "H_B", and the depth of the sinking surface is defined as parameter "H_A", respectively. After a series of test with varies loading, the relationship of the four parameters can be summarized as Eq. (1). Actually, in the field the distance of the boundaries "D" may be very far because the real domain is a half-infinite space. Therefore "D" will be much larger than the diameter of the chamber "d" as the Eq. (2). Under the condition, the depth of the sinking surface "H_A" will become very small even close to zero as the Eq. (3). In the field it should be difficult to observed. However, back to the Eqs. (1) and (2), since "D" will be much larger than "d", the height of the floater "H_B" must be larger than zero as the Eq. (4). In practice, it means the Tubing-Floater Device proposed herein will be a useful tool to detect the mud-pumping phenomenon, even for a light case.

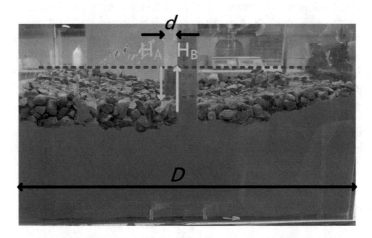

Fig. 11. The sectional profile after mud-pumping generating

An automatic recording Tubing-Floater Device is now developing. The function of that will be expected to log the real-time data as shown in Fig. 12. The estimated tendency of the mud-pumping behaviour can be more accurately obtained.

$$H_A \cdot D \propto H_B \cdot d \tag{1}$$

$$D \gg d \tag{2}$$

$$H_A \to 0 \tag{3}$$

$$H_B > 0 \tag{4}$$

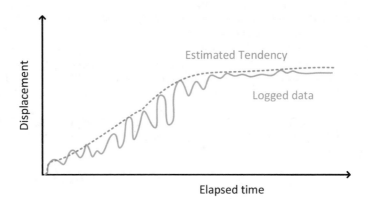

Fig. 12. Schematic diagram illustrating the logged data by automatic recording Tubing-Floater Device in the future

5   Remarks

In this paper, the designs of Tubing-Floater Device for detecting railway mud-pumping have been proposed. According to a series of laboratorial tests, the function of the device is proof. The device can work well and detect light to heavy mud-pumping. It will be helpful the field work to investigate the mud-pumping distribution and find hot zones by continuously monitoring. The maintenance of the railway foundation will be more simplified to evaluate the area and depth that the mud already pumped. Moreover, the hazard maybe raised by this phenomenon can be prevent as well.

Acknowledgments. Financial support from Ministry of Science and Technology of Taiwan (MOST 104-2622-M-159-001 -CC3, 105-2622-M-159 -001 -CC3) is very appreciated. Test area was provided by Taiwan Railways Administration. Their kindly helps are very much appreciated.

References

Katsumasa, Y.: Mud pumping on tracks—present state and countermeasures. Jpn. Railw. Eng. **17** (4), 20–21 (1978)

Ayres, D.J.: Geotextiles or geomembranes in track? British railway experience. Geotextile Geomembr. **3**(2–3), 129–142 (1986)

Alobaidi, I., Hoare, D.J.: Factors affecting the pumping of fines at the subgrade- subbase interface of highway pavements: a laboratory study. Geosynth. Int. **1**(2), 221–259 (1994)

Alobaidi, I., Hoare, D.J.: The development of pore water pressure at the subgrade-subbase interface of a highway pavement and its effect on pumping of fines. Geotext. Geomembr. **14** (2), 111–135 (1996)

Takatoshi, I.: Measure for the Stabilization of Railway Earth Structure. Japan Railway Technical Service, Tokyo, Japan (1997)

Indraratna, B., Ionescu, D., Christie, D., Chowdhurry, R.: Compression and degradation of railway ballast under one-dimensional loading. Austr. Geomech. J. **12**, 48–61 (1997)

Aw, E.S.: Novel monitoring system to diagnose rail track foundation problems. M.S. thesis. Massachusetts Institute of Technology, Cambridge, MA (2004)

Ghataora, G.S., Burns, B., Burrow, M.P.N., Evdorides, H.T.: Development of an index test for assessing anti- pumping materials in railway track foundations. In: Proceedings of the First International Conference on Railway Foundations, Railfound 2006. University of Birmingham, Birmingham, UK, pp. 355–366 (2006)

ASTM Standard D698: Standard Test Methods for Laboratory Compaction Characteristics of Soil Using Standard Effort, ASTM International, West Conshohocken, PA (2007)

Aw, E.S.: Low cost monitoring system to diagnose problematic rail bed: case study at mud pumping site. Ph.D. dissertation. Massachusetts Institute of Technology, Cambridge, MA (2007)

Duong, T.V., Cui, Y.J., Tang, A.M., Dupla, J.C., Canou, J., Calon, N., Robinet, A., Chabot, B., De Laure, E.: Physical model for studying the migration of fine particles in the railway substructure. Geotech. Test. J. ASTM **37**(5), 1–12 (2014)

Author Index

© Springer Nature Switzerland AG 2019
S. El-Badawy and J. Valentin (Eds.): GeoMEast 2018, SUCI, pp. 161–162, 2019.
https://doi.org/10.1007/978-3-030-01911-2

Printed in the United States
By Bookmasters